Standards and Guidelines for Electroplated Plastics

Standards and Guidelines

4th edition

Standards and Guidelines for Electroplated Plastics

American Society
of Electroplated Plastics

Prentice-Hall, Inc., Englewood Cliffs, N.J. 07623

Library of Congress Cataloging in Publication Data
Main entry under title:

Standards and guidelines for electroplated plastics.

Rev. ed. of: Standards and guidelines. c1979.
Bibliography: p.
1. Metal coating. 2. Plastic coatings.
I. American Society of Electroplated Plastics.
TP1175.M4S73 1983 668.4′19 83-3129
ISBN 0-13-842310-5

Editorial/production supervision and
interior design: *Gretchen K. Chenenko*
Cover design: *Edsal Enterprises*
Manufacturing buyer: *Anthony Caruso*

Printed in the United States of America

10 9 8 7 6 5 4 3 2 1

ISBN 0-13-842310-5

Prentice-Hall International, Inc., *London*
Prentice-Hall of Australia Pty. Limited, *Sydney*
Editora Prentice-Hall do Brasil, Ltda., *Rio de Janeiro*
Prentice-Hall Canada Inc., *Toronto*
Prentice-Hall of India Private Limited, *New Delhi*
Prentice-Hall of Japan, Inc., *Tokyo*
Prentice-Hall of Southeast Asia Pte. Ltd., *Singapore*
Whitehall Books Limited, *Wellington, New Zealand*

Contents

Preface

This is the fourth edition of the Standards and Guidelines first published by the American Society of Electroplated Plastics in 1972. This edition has been revised to incorporate the latest changes that have taken place in the technology of electroplating on plastics since the publication of the third edition of the Guidelines in December, 1984.

The purpose of these Standards and Guidelines is to ensure the quality and durability of electroplated plastic articles. It is ASEP's hope that emphasis on quality will continue to foster the growth of electroplating plastics.

The American Society of Electroplated Plastics is deeply indebted to the Technical Advisory Group and to the member company representatives that so graciously gave of their time and talents to generate and review this work.

About ASEP

The American Society of Electroplated Plastics was organized in 1966 and incorporated in 1967 in Pennsylvania as a nonprofit organization. Since its inception, the Society has expanded its membership throughout the world. More than 100 member companies have participated in its important work. ASEP is the only national society representing the electroplated plastics industry in the United States.

The basic aim of the Society is to improve the efficiency of the industry and to promote the wider use of electroplated plastics. Both are essential to the industry's continued growth and prosperity.

ASEP strives to: (1) promote the development, interest, and general welfare of the electroplaters, molders, suppliers, and customers of the products used in the industry, and to safeguard their common business interests; (2) collect, preserve, and disseminate to the members, the industry, and general public information as to the nature, importance, and methods of the industry; (3) develop standards of quality for the manufacture and electroplating of articles handled by metal finishing firms; (4) appear on behalf of the Society and its members before legislative committees, governmental departments and agencies, and other bodies, to cooperate on such matters affecting the industry, and to promote in all lawful ways the interests and welfare of the industry.

Introduction

Sections in the ASEP Standards and Guidelines will discuss in detail proper material selection, part design, tool design, processing, and evaluation of the finished part. This discussion is aimed at producing parts that are of consistent good quality and are reliable in their end use applications.

Materials discussed in this book are either in use commercially in electroplated plastic parts or have been named by their manufacturers as being plateable.

One can examine the physical properties of the plastic materials now available for plating. After determining in general whether plastic will provide the necessary property requirements, one must select the proper substrate to provide the required service in the plated part. Reviewed here are the general performance qualifications of the various substrates when plated, and the typical applications for each.

Because plastics are nonconductive, the preparation for electroplating requires a series of steps that results in the chemical deposition of a thin conductive metal film on the plastic surface. Once the plastic has been made conductive, it may be electroplated with a metal or combination of metals for decorative or functional purposes. Each step in the preplating cycle must be carefully controlled to obtain acceptable and results.

The quality and performance of an electroplated plastic part are also influenced by the processing that takes place prior to preplating. A knowledge of the properties of plateable plastic materials is necessary to select the correct plastic for a particular application. The design of the part and the mold must be made with electroplating in mind. With two-phase materials like ABS, there is a close relationship between injection molding parameters and the adhesion of subsequently applied metal coatings. These topics are discussed in this edition. Test methods and procedures for evaluating the performance of electroplated plastics are also discussed in detail.

1

Properties of Plateable Plastic Materials

1.1 PLATING GRADE ABS

ABS plastic has the highest volume usage for commercial plating applications. The chemistry of the resin lends itself favorably to the processing techniques used to obtain smooth adherent metal deposits. The design versatility and ease of processability of ABS plastic makes it an ideal material for economically producing a variety of durable products for a broad range of consumer markets.

The first plastic substrate to be used in volume quantities for electroplating was general-purpose ABS. General-purpose ABS is suitable for a variety of functional and decorative applications, such as small appliance housings, control knobs, dials, and switches.

The highest-volume ABS usage is in exterior automotive parts, which require a higher-performance grade of ABS.

As the applications for plated plastics grew, the size and complexity of the part to be plated also grew. This made adhesion of plate to plastic and thermal cycle requirements increasingly difficult. For larger parts, the expansion and contraction characteristics of the plastic substrate become critical. As the service temperature changes, the dimensional change between the plastic substrate and the plate deposit increases stresses between the plate and the plastic. This is due to thermal coefficient disparities of the plastic and the plated metal. In some extremes the plate can fail by cracking.

To meet industry needs, resin suppliers developed higher-performance grades of ABS for electroplating. These overcome many difficulties encountered with larger and more complex-shaped parts. The higher-performance grades of ABS plastics withstand greater temperature extremes. The high-performance grades of ABS also offer lower coefficient of linear thermal expansion of the plastic substrate. This helps large and complex-shaped parts to meet thermal cycle requirements.

Typical applications of the high-performance grades of ABS are automotive exterior trim applications such as radiator grilles, headlamp bezels, tail-lamp housings, wheel covers, license frames, and pillar moldings. These grades are also used for marine hardware and plumbing fixtures.

1.2 ABS-POLYCARBONATE ALLOYS

Building on the in-house capability and capacity that exists for plating ABS substrates, the ABS-polycarbonate substrates extend the environmental performance capabilities of ABS substrates with only minor modifications of plating systems.

Electroplated ABS-polycarbonate parts have significantly higher part warpage and thermal cycle capabilities than even high-performance ABS substrates. At temperatures greater than 180°F, this substrate, when plated with a standard electroplate, retains a greater percentage of its room-temperature design and strength properties than does ABS. Also, in a strain situation, the electroplated ABS-polycarbonate alloy composite tends to resist the onset of metal crack initiation until at or beyond the yield point of the composite, making it a more suitable substrate for slow strain rate load-bearing applications than ABS. Creep of the electroplated alloy at stress levels of 1000 psi and temperatures of 180°F is minimal.

As a result of these performance extensions of standard ABS capabilities, the ABS-polycarbonate alloy could be considered for load-functional applications and heat environments up to 230°F. In addition, with proper part design this electroplated substrate could be a candidate for applications required to meet UL-945V flame test standards.

Major application usage includes exterior automotive trim where high heat resistance is required, major appliance trim such as range end caps and oven door escutcheons and handles, and institutional lighting. In the plumbing industry, applications requiring contact with water above 160°F and parts that are difficult to gain Canadian Standards Association (CSA) thermal cycle approval in high-performance ABS substrates should be evaluated in this alloy.

1.3 MINERAL-REINFORCED NYLON (MRN)

Mineral-reinforced nylon offers a unique mix of properties for a plateable plastic. Mineral-reinforced nylons have outstanding strength, rigidity, and toughness.

They also have very high distortion temperatures as well as a low thermal-expansion coefficient. These aid in meeting thermal cycle and strength requirements for a wide range of part applications.

Mineral-reinforced nylon is suitable for components subjected to moderate heat in service, such as oven control knobs and levers and in automotive headlamp housings. The strength, rigidity, and impact resistance of MRN are important for use in functional applications such as automotive window cranks, door handles, wheel covers, and mirror housings. MRN is also suitable for non-water bearing applications in plumbing fixtures, household hardware, and marine hardware.

1.4 MODIFIED POLYPHENYLENE OXIDE

Modified polyphenylene oxides (PPO) comprise a family of engineering thermoplastics offering a wide range of desirable design properties. Plateable grades of modified PPO find use in a number of different markets, including automotive, appliances, and plumbing.

The key properties of plateable PPO are a low coefficient of linear thermal expansion, excellent low-temperature impact, and a high distortion temperature. The inherent dimensional stability of modified polyphenylene oxide over wide temperature extremes aids in withstanding thermal cycle testing. Modified PPO offers a slightly higher heat deflection temperature than ABS and better low-temperature impact. That makes it a suitable substrate for certain high- and low-temperature applications.

Typical automotive applications of plated PPO include exterior trim such as radiator grilles and wheel discs and hubs. In the appliance market, this material has been used for range, washer, and dryer back panels. Various plumbing fixtures, such as faucet handles, tub spouts, and shower heads, have been made from plateable PPO-based material.

1.5 POLYESTERS

Polyester and vinyl ester resins embrace a family of engineering thermoset systems that present a new, wide range of physical properties. These resin systems are usually reinforced with fiberglass, mineral, or mineral glass combinations. The plateable grades of polyester and vinyl ester resins provide opportunities in the automotive, plumbing, appliance, business machine, cosmetics, and consumer products markets.

The inherent dimensional stability of these resins helps in withstanding thermal cycle testing. These systems offer opportunities for the future and will require the latest in innovative plating techniques. These resins offer excellent chemical resistance, low moisture absorption, and high heat resistance.

1.6 POLYSULFONE

Polysulfone has been used successfully in applications where good heat resistance and good dimensional stability of the part are required. Hot-water plumbing components and appliance parts subject to high-heat environments are typical applications. Filled polysulfone has been successfully used for decoratively plated camera housings, where the dimensional stability over wide temperature extremes helps to maintain necessary tolerance for precision devices.

1.7 RESIN PROPERTIES

Many plastic manufacturers have indicated that they have plating-grade polymers in their product line. The physical properties of some of these materials are listed in this section for the convenience of those specifying plated plastics. A summary of the physical properties listed is shown in Table 1.1. This includes the property,

TABLE 1.1

Properties	Condition	Units	Test Method
Izod Impact	23°C (73°F)	J/m (ft-lbs/in)	D-256
3.2mm (1/8-in)	−40°C (−40°F)	J/m (ft-lbs/in)	Method A
Bar, Notched			
Tensile Strength	23°C (73°F)	MPa (psi)	D-638
3.2mm (1/8-in) Bar	71°C (160°F)	MPa (psi)	
Tensile Modulus	23°C (73°F)	MPa (psi)	D-638
Flexural Strength	23°C (73°F)	MPa (psi)	D-790
	71°C (160°F)	MPa (psi)	
	−40°C (−40°F)	MPa (psi)	
Flexural Modulus	23°C (73°F)	MPa (psi)	D-790
	71°C (160°F)	MPa (psi)	
	−40°C (−40°F)	MPa (psi)	
Hardness	23°C (73°F)	Rockwell	D785
Coefficient of		cm/cm°C × 10^{-5}	D696
Thermal Expansion		(in/in°F × 10^{-5})	
Specific Gravity			D792
			Method A
Deflection Temperature			
Unannealed	1820 kPa (264 psi)	°C (°F) @	D-648
	455 kPa (66 psi)	0.25mm (0.01 in)	
		deflection	
Annealed	1820 kPa (264 psi)		
	455 kPa (66 psi)		
Mold Shrinkage		cm/cm (in/in)	D-790

the specimen size, the condition of the test, the units of measurement, and the ASTM test method used.

The physical property data shown in Tables 1.2–1.13 is meant only as a guide to material users. Specimen size and molding conditions may vary from supplier to supplier. Since these parameters have an effect on the value obtained for a given test method, individual suppliers should be contacted for additional information when making a material selection.

TABLE 1.2 Plating-Grade Polymers

	Generic Type: ABS Manufacturer: GE Plastics Trade Name: Cyclolac Grade: EP-3510	
Properties	Condition	
Izod Impact	23°C (73°F)	374 J/m (7.0 ft-lbs/in)
	−40°C (−40°F)	69 J/m (1.3 ft-lbs/in)
Tensile Strength	23°C (73°F)	38 MPa (5500 psi)
	71°C (160°F)	19 MPa (2800 psi)
Tensile Modulus	23°C (73°F)	2300 MPa (300,000 psi)
Flexural Strength	23°C (73°F)	69 MPa (10,000 psi)
	71°C (160°F)	33 MPa (4800 psi)
	−40°C (−40°F)	105 MPa (15,300 psi)
Flexural Modulus	23°C (73°F)	2300 MPa (340,000 psi)
	71°C (160°F)	1400 MPa (200,000 psi)
	−40°C (−40°F)	2700 MPa (390,000 psi)
Hardness	23°C (73°F)	R105
Coefficient of Linear Thermal Expansion		6.8×10^{-5} cm/cm/°C (3.8×10^{-5} in/in/°F)
Specific Gravity		1.04
Deflection Temperature		
Unannealed	1820kPa (264 psi)	87°C (188°F)
	455kPa (66 psi)	99°C (210°F)
Annealed	1820kPa (264 psi)	100°C (212°F)
	455kPa (66 psi)	102°C (215°F)
Mold Shrinkage	Range	.005–.008 cm/cm (in/in)

TABLE 1.3 Plating-Grade Polymers

	Generic Type: ABS Manufacturer: GE Plastics Trade Name: Cyclolac Grade: EPB-3570	
Properties	Condition	
Izod Impact	23°C (73°F)	267 J/m (5.0 ft-lbs/in)
	−40°C (−40°F)	64 J/m (1.2 ft-lbs/in)
Tensile Strength	23°C (73°F)	44 MPa (7500 psi)
	71°C (160°F)	25 MPa (3600 psi)
Tensile Modulus	23°C (73°F)	2600 MPa (360,000 psi)
Flexural Strength	23°C (73°F)	80 MPa (13,000 psi)
	71°C (160°F)	37 MPa (5400 psi)
	−40°C (−40°F)	133 MPa (19,300 psi)
Flexural Modulus	23°C (73°F)	2700 MPa (390,000 psi)
	71°C (160°F)	1700 MPa (250,000 psi)
	−40°C (−40°F)	3000 MPa (440,000 psi)
Hardness	23°C (73°F)	R112
Coefficient of Linear Thermal Expansion		6.5×10^{-5} cm/cm/°C (3.6×10^{-5} in/in/°F)
Specific Gravity		1.05
Deflection Temperature		
Unannealed	1820kPa (264 psi)	96°C (203°F)
	455kPa (66 psi)	103°C (217°F)
Annealed	1820kPa (264 psi)	100°C (212°F)
	455kPa (66 psi)	104°C (220°F)
Mold Shrinkage	Range	.005–.007 cm/cm (in/in)

TABLE 1.4 Plating-Grade Polymers

	Generic Type: ABS Manufacturer: GE Plastics Trade Name: Cycolac Grade: CKM1	
Properties	Condition	
Izod Impact	23°C (73°F)	320 J/m (3.5 ft-lbs/in)
	−40°C (−40°F)	1.0 ft-lbs/in
Tensile Strength	23°C (73°F)	43 MPa (5200 psi)
	71°C (160°F)	—
Tensile Modulus	23°C (73°F)	2200 MPa (280,000 psi)
Flexural Strength	23°C (73°F)	76 MPa (9100 psi)
	71°C (160°F)	—
	−40°C (−40°F)	114 MPa (16,500 psi)
Flexural Modulus	23°C (73°F)	2400 MPa (330,000 psi)
	71°C (160°F)	—
	−40°C (−40°F)	2900 MPa (420,000 psi)
Hardness	23°C (73°F)	R97
Coefficient of Linear Thermal Expansion		10.8×10^{-5} cm/cm/°C (6.0×10^{-5} in/in/°F)
Specific Gravity		1.23
Deflection Temperature		
Unannealed	1820kPa (264 psi)	103°C (217°F)
	455kPa (66 psi)	96°C (205°F)
Annealed	1820kPa (264 psi)	93°C (200°F)
	455kPa (66 psi)	101°C (215°F)
Mold Shrinkage		.007–.008 cm/cm (in/in)

TABLE 1.5 Plating-Grade Polymers

	Generic Type: ABS Manufacturer: GE Plastics Trade Name: Cycolac Grade: KJW	
Properties	Condition	
Izod Impact	23°C (73°F)	267 J/m (4.0 ft-lbs/in)
	−40°C (−40°F)	—
Tensile Strength	23°C (73°F)	41 MPa (6000 psi)
	71°C (160°F)	—
Tensile Modulus	23°C (73°F)	2100 MPa (300,000 psi)
Flexural Strength	23°C (73°F)	69 MPa (10,000 psi)
	71°C (160°F)	—
	−40°C (−40°F)	103 MPa (15,000 psi)
Flexural Modulus	23°C (73°F)	2300 MPa (330,000 psi)
	71°C (160°F)	—
	−40°C (−40°F)	2600 MPa (380,000 psi)
Hardness	23°C (73°F)	R99
Coefficient of Linear Thermal Expansion		20.2×10^{-5} cm/cm/°C (11.2×10^{-5} in/in/°F)
Specific Gravity		1.23
Deflection Temperature		
Unannealed	1820kPa (264 psi)	88°C (190°F)
	455kPa (66 psi)	93°C (200°F)
Annealed	1820kPa (264 psi)	96°C (205°F)
	455kPa (66 psi)	99°C (210°F)
Mold Shrinkage		.005–.008 cm/cm (in/in)

TABLE 1.6 Plating-Grade Polymers

	Generic Type:	ABS-Polycarbonate Alloy
	Manufacturer:	GE Plastics
	Trade Name:	Cycoloy
	Grade:	EHA-3522
Properties	**Condition**	
Izod Impact	23°C (73°F)	342 J/m (6.9 ft-lbs/in)
	−29°C (−20°F)	85 J/m (1.2 ft-lbs/in)
	−40°C (−40°F)	37 J/m (0.7 ft-lbs/in)
Tensile Strength	23°C (73°F)	50 MPa (6600 psi)
	71°C (160°F)	26 MPa (3750 psi)
Tensile Modulus	23°C (73°F)	2400 MPa (350,000 psi)
Flexural Strength	23°C (73°F)	83 MPa (11,400 psi)
	71°C (160°F)	46 MPa (6600 psi)
	−40°C (−40°F)	117 MPa (17,000 psi)
Flexural Modulus	23°C (73°F)	2500 MPa (370,000 psi)
	71°C (160°F)	1800 MPa (260,000 psi)
	−40°C (−40°F)	2900 MPa (420,000 psi)
Hardness	23°C (73°F)	R111
Coefficient of Linear Thermal Expansion		6.2×10^{-5} cm/cm/°C (3.45×10^{-5} in/in/°F)
Specific Gravity		1.09
Deflection Temperature		
Unannealed	1820kPa (264 psi)	110°C (230°F)
	455kPa (66 psi)	119°C (247°F)
Annealed	1820kPa (264 psi)	121°C (249°F)
	455kPa (66 psi)	124°C (256°F)
Mold Shrinkage	Range	.005–.008 cm/cm (in/in)

TABLE 1.7 Plating-Grade Polymers

Generic Type: Nylon 6
Manufacturer: Allied Chemical
Trade Name: Capron
Grade: 8260

Properties	Condition	
Izod Impact	23°C (73°F)	—
	−40°C (−40°F)	—
Tensile Strength	23°C (73°F)	83 MPa (13,000 psi)
	71°C (160°F)	—
Tensile Modulus	23°C (73°F)	—
Flexural Strength	23°C (73°F)	145 MPa (21,000 psi)
	71°C (160°F)	—
	−40°C (−40°F)	—
Flexural Modulus	23°C (73°F)	5500 MPa (800,000 psi)
	71°C (160°F)	—
	−40°C (−40°F)	—
Hardness	23°C (73°F)	—
Coefficient of Linear Thermal Expansion		5.0×10^{-5} cm/cm/°C (2.8×10^{-5} in/in/°F)
Specific Gravity		1.47
Deflection Temperature		
Unannealed	1820kPa (264 psi)	124°C (255°F)
	455kPa (66 psi)	—
Annealed	1820kPa (264 psi)	—
	455kPa (66 psi)	—
Mold Shrinkage	Range	.009–.011 cm/cm (in/in)

TABLE 1.8

Generic Type:	Polyphthalamide
Manufacturer:	Amoco Performance Plastics
Trade Name:	AMODEL
Grade:	A-1240

Properties	Condition	
Izod Impact	23°C (73°F)	0.8 ft-lbs/in
Tensile Strength	23°C (73°F)	17,000 psi
Tensile Modulus	—	—
Flexural Strength	23°C (73°F)	30,000 psi
Flexural Modulus	23°C (73°F)	1,300,000 psi
Hardness	23°C (73°F)	R125
Coefficient of Linear		1.9×10^{-5} cm/cm/°C
Thermal Expansion[1]		(3.4×10^{-5} in/in/°F)
Specific Gravity		1.54
Deflection Temperature		
Unannealed	—	—
Annealed	264 psi	183°C (361°F)
	66 psi	282°C (540°F)
Mold Shrinkage		0.008 in/in

[1]Test method E-831

TABLE 1.9 Plating-Grade Polymers

Generic Type: Polyphenylene Oxide (Modified)
Manufacturer: GE Plastics
Trade Name: Noryl
Grade: PN-235

Properties	Condition	
Izod Impact	23°C (73°F)	267 J/m (6.5 ft-lbs/in)
	−40°C (−40°F)	2.5 ft-lbs/in
Tensile Strength	23°C (73°F)	48.3 MPa (5000 psi)
	71°C (160°F)	—
Tensile Modulus	23°C (73°F)	—
Flexural Strength	23°C (73°F)	67.6 MPa (8000 psi)
	71°C (160°F)	—
	−40°C (−40°F)	—
Flexural Modulus	23°C (73°F)	2069 MPa (260,000 psi)
	71°C (160°F)	—
	−40°C (−40°F)	—
Hardness	23°C (73°F)	
Coefficient of Linear Thermal Expansion		7.6×10^{-5} cm/cm/°C (4.2×10^{-5} in/in°F)
Specific Gravity		1.05
Deflection Temperature		
Unannealed	1820kPa (264 psi)	107°C (225°F)
	455kPa (66 psi)	113°C (235°F)
Annealed	1820kPa (264 psi)	—
	455kPa (66 psi)	—
Mold Shrinkage		.006 cm/cm (in/in)

TABLE 1.10 Plating-Grade Polymers

	Generic Type: Polyester Manufacturer: GE Plastics Trade Name: VALOX Grade: 745	
Properties	Condition	
Izod Impact	23°C (73°F)	320 J/m (1.5 ft-lbs/in)
	−40°C (−40°F)	0.7 ft-lbs/in
Tensile Strength	23°C (73°F)	43 MPa (6800 psi)
	71°C (160°F)	—
Tensile Modulus	23°C (73°F)	2200 MPa (551,000 psi)
Flexural Strength	23°C (73°F)	76 MPa (13,000 psi)
	71°C (160°F)	—
	−40°C (−40°F)	114 MPa (16,500 psi)
Flexural Modulus	23°C (73°F)	2400 MPa (500,000 psi)
	71°C (160°F)	—
	−40°C (−40°F)	2900 MPa (420,000 psi)
Hardness	23°C (73°F)	R112
Coefficient of Linear Thermal Expansion		7.6×10^{-5} cm/cm/°C (4.2×10^{-5} in/in/°F)
Specific Gravity		1.46
Deflection Temperature		
Unannealed	1820kPa (264 psi)	88°C (190°F)
	455kPa (66 psi)	160°C (320°F)
Annealed	1820kPa (264 psi)	99°C (210°F)
	455kPa (66 psi)	101°C (215°F)
Mold Shrinkage		.008–.010 cm/cm (in/in)

25% mineral reinforced

TABLE 1.11 Plating-Grade Polymers

	Generic Type: Polyester	
	Manufacturer: GE Plastics	
	Trade Name: Heavy Valox	
	Grade: HV 7065	
Properties	Condition	
Izod Impact		
Notched	23°C (73°F)	320 J/m (0.6 ft-lbs/in)
Unnotched	−40°C (−40°F)	3.0 ft-lbs/in
Tensile Strength	23°C (73°F)	43 MPa (6500 psi)
	71°C (160°F)	—
Tensile Modulus	23°C (73°F)	2200 MPa (320,000 psi)
Flexural Strength	23°C (73°F)	76 MPa (13,000 psi)
	71°C (160°F)	—
	−40°C (−40°F)	114 MPa (16,500 psi)
Flexural Modulus	23°C (73°F)	2400 MPa (760,000 psi)
	71°C (160°F)	—
	−40°C (−40°F)	2900 MPa (420,000 psi)
Hardness	23°C (73°F)	R115
Coefficient of Linear Thermal Expansion		8.5×10^{-5} cm/cm/°C (4.7×10^{-5} in/in/°F)
Specific Gravity		2.40
Deflection Temperature		
Unannealed	1820kPa (264 psi)	90°C (195°F)
	455kPa (66 psi)	196°C (385°F)
Annealed	1820kPa (264 psi)	99°C (210°F)
	455kPa (66 psi)	101°C (215°F)
Mold Shrinkage		.006 cm/cm (in/in)

65% mineral reinforced PBT blend

TABLE 1.12 Plating-Grade Polymers

	Generic Type: Polysulfone Manufacturer: Union Carbide Trade Name: Udel Grade: P-6050	
Properties	**Condition**	
Izod Impact	23°C (73°F)	80 J/m (1.5 ft-lbs/in)
	−40°C (−40°F)	—
Tensile Strength	23°C (73°F)	90 MPa (13,000 psi)
	71°C (160°F)	—
Tensile Modulus	23°C (73°F)	5000 MPa (720,000 psi)
Flexural Strength	23°C (73°F)	138 MPa (20,000 psi)
	71°C (160°F)	—
	−40°C (−40°F)	—
Flexural Modulus	23°C (73°F)	5200 MPa (750,000 psi)
	71°C (160°F)	—
	−40°C (−40°F)	—
Hardness	23°C (73°F)	M75
Coefficient of Linear Thermal Expansion		4.5×10^{-5} cm/cm/°C (2.5×10^{-5} in/in/°F)
Specific Gravity		1.40
Deflection Temperature		
Unannealed	1820kPa (264 psi)	179°C (355°F)
	455kPa (66 psi)	—
Annealed	1820kPa (264 psi)	—
	455kPa (66 psi)	—
Mold Shrinkage		.005 cm/cm (in/in)

TABLE 1.13 Plating-Grade Polymers

Generic Type: Modified Polysulfone
Manufacturer: Amoco
Trade Name: Mindel
Grade: A-670

Properties	Condition	
Izod Impact	23°C (73°F)	374 J/m (7.0 ft-lb/in)
	−40°C (−40°F)	—
Tensile Strength	23°C (73°F)	49 MPa (7,200 psi)
	−40°C (−40°F)	—
Tensile Modulus	23°C (73°F)	2123 MPa (308,000 psi)
Flexural Strength	23°C (73°F)	83 MPa (12,000 psi)
Flexural Modulus	23°C (73°F)	2180 MPa (316,000 psi)
	71°C (160°F)	—
	−40°C (−40°F)	—
Hardness	23°C (73°F)	
Coefficient of Linear Thermal Expansion		6.5×10^{-5} mm/mm/°C (3.6×10^{-5} in/in/°F)
Specific Gravity		1.13
Deflection Temperature		
Unannealed	1820kPa (264 psi)	149°C (300°F)
	455kPa (66 psi)	—
Annealed	1820kPa (264 psi)	—
	455kPa (66 psi)	—
Mold Shrinkage		0.007 mm/mm (in/in)

2

Part Design Considerations for Electroplated Plastics

A guide for the design engineer

2.1 ELECTROPLATED PLASTICS

Electroplated plastics, well designed and properly applied to take advantage of the unique properties of plastic, are not only economical but also superior in quality to conventional materials that have been traditionally used for many years. Their potential use is limited only by the imagination. Their success and growth are a substantial economic factor in American industry and are limited only by the care and attention the design engineer and stylist give to application and design.

Final economical quality of a product can be no better than the tool used to produce it, and the basic consideration is design. Once the decision has been made to use plastics, by following a few design rules and establishing realistic product specifications, a superior product will be the result.

Furthermore, not only is a well-designed part essential to the successful application of the part, it is also essential to the molder, the plater, and the end user, for they all must be profitable as a result of the application.

Therefore, the elements of proper part design must be known to all concerned.

To assure a successful application, the final part design should be arrived at by a logical sequence of steps. Table 2.1 lists a plated plastic design procedure that could be used for arriving at the final part design. The sequence of the design evolution may vary, but all steps should appear during design and introduction of the plated-plastic application.

TABLE 2.1 Plated-Plastic Design Procedure

1. Appraisal of Application (See Product Specifications)
    - Set Performance Requirements
    - Set Performance Specifications
2. Design Proposal
    - Basic Engineering and Industrial Design
    - Economic Analysis
3. Prototype Design
    - Prototype Mold Design
4. Fabricate Prototype
    - Resin Selection
    - Preplate Selection
    - Electroplate Composite Determination
5. Parts Testing—Test to Performance Specifications Set in 1
    - Accelerated Laboratory Tests
    - In-Use or Field Tests
6. Evaluation of Test Results
7. Final Design
    - Functional
    - Aesthetic
8. Specifications
    - Vendor Selection
    - Tooling Procurement
9. Part Production
    - Continued Part Qualification
    - Part Performance Analysis

TABLE 2.2 Part Design Consideration

1. Molded Plastics Design
2. Appearance Design
3. Electroplating Distribution
4. Thermal Performance Requirements
5. Plate Corrosion Performance

Unfortunately, the "perfect" part design does not guarantee a successful plated-plastic application. This point is verified by the presence of the subsequent sections of the ASEP Guidelines. However, the proper use of the material offered in this section will start the application in the right direction.

The outline in Table 2.2 shows the breakdown of plated-plastic part design considerations used in this chapter. Reference to the outline will be helpful in locating specific design information. Basic considerations are listed under each design characteristic shown in Table 2.2.

Review of the topics listed in the part design considerations outline may lead the reader to wonder what relevance some of the topics have to the actual part design.

The answer is that when—and only when—the design engineer understands the interplay of all facets of plating on plastics will full potential of plated plastics be realized.

2.2 MOLDED PLASTIC DESIGN

Properly designing an electroplated plastic part starts with the design of the plastic molding. Basic design recommendations made here are intended to aid the designer in avoiding design errors that would adversely affect the quality of electroplated plastic parts. Additional design information is available from most plastic resin suppliers.

2.2.1 Nominal Wall Thickness

A completely uniform wall thickness is ideal but rarely practical. Uniform wall thickness throughout a part promotes uniform cooling and minimized warpage and shrinkage defects.

Nominal wall thickness, illustrated in Figure 2.1, is the overall design thickness of most of all sections of a part.

Nominal wall thickness should be in the range of 2.3–3.0 mm (0.09–0.12 in). Maximum wall thickness should not exceed 3.8 mm (0.15 in) because of cooling problems and the tendency of heavier sections to exhibit sink marks after

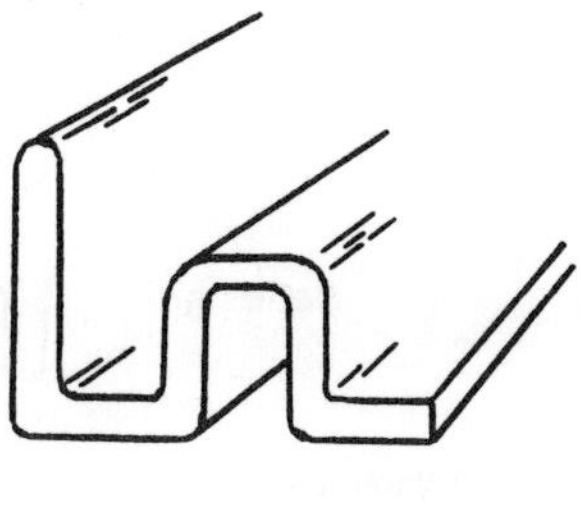

Figure 2.1 Nominal thickness should be maintained throughout the part.

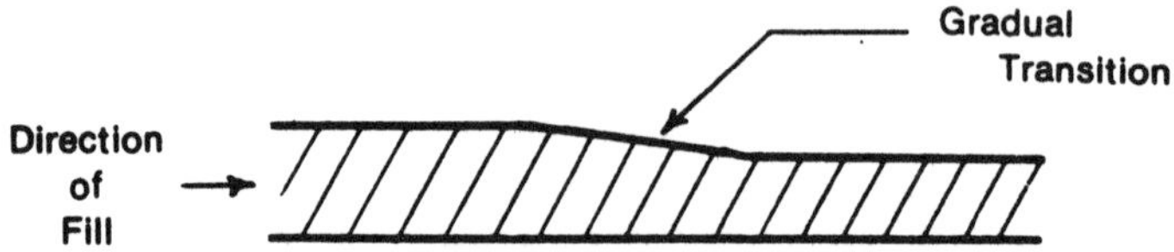

Figure 2.2 Wall thickness transitions

cooling. The minimum wall thickness should not be less than 1.9 mm (.075 in) because thinner sections will be weak and may be difficult to properly fill in the molding process.

By designing hollow ribs, bosses, and elevation changes, intricate part geometry can be achieved while maintaining nominal thickness throughout the part. Proper use of these concepts will minimize warpage, save molding material, and save cycle time in the molding operation.

Thickness changes should be avoided whenever possible because the resultant uneven cooling will cause shrinkage and warping of the part. When wall thickness variations are necessary, they should be kept to a minimum, and transitions should be gradual and as uniform as possible.

In general, if wall thickness variations are necessary in a part, the part should fill in the molding cycle from the heavier section to the thinner section. Trying to fill the part from the thinner section to the heavier section tends to cause stress and material-orientation problems in the thinner sections. This is shown in Figure 2.2.

Large weight-to-surface ratios should definitely be avoided. Figures 2.3 and 2.4 illustrate the redesign of two solid plastic knobs which, in the solid form, would have had unacceptable shrinkage characteristics in the molding as well as serious thermal cycle problems after plating.

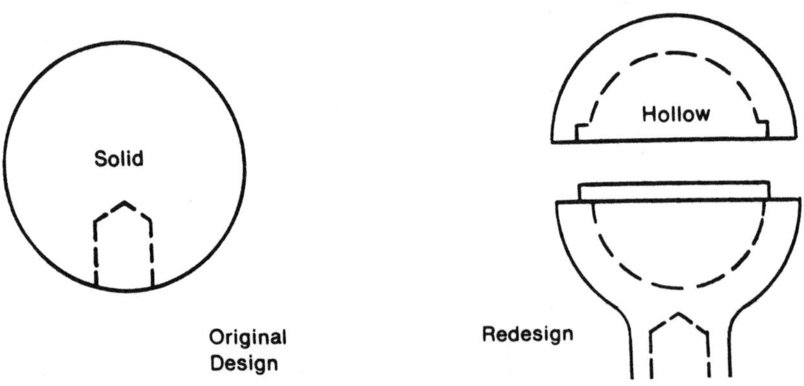

Figure 2.3 Gear-shift knob redesign

Figure 2.4 Lever knob redesign

2.2.2 Ribs

Ribs are linear wall projections from adjacent plane surfaces. Ribs are used to reduce the bulk or weight of parts while meeting high strength and rigidity requirements. Interconnecting ribs are used in many cases to prevent bowing or warpage of large plane surfaces. An example is shown in Figure 2.5.

Ribs are also used extensively in boss design to allow minimum boss wall thickness without sacrificing strength. An example is shown in Figure 2.6.

Ribs should be designed so that they are 50% to 60% of the supported wall thickness at their point of intersection. A number of smaller ribs may be required instead of one large rib in order to maintain this wall thickness relationship. This wall thickness ratio is necessary to avoid visible sink marks on the opposite side of the supported wall from the rib. This relationship is illustrated in Figure 2.7.

Figure 2.5 Reinforcing ribs

Figure 2.6 Rib support for bosses

Ribs should always be designed with proper radii at their wall intersection to avoid stress concentration and with a minimum of 1° draft for ease of removal from the mold.

A typical use of a rib for wall support is shown in Figure 2.8.

2.2.3 Bosses

Bosses are projections from the part surface that provide for attachment and support of related components. They may be hollow or solid. Solid bosses are sometimes called studs.

The wall thickness recommendations for ribs also apply to bosses. Boss wall thicknesses that exceed 50% of the intersecting wall thickness will tend to cause visible sink marks on the part surface. Ribs should be added to the boss for strength rather than increasing its wall thickness above recommended levels.

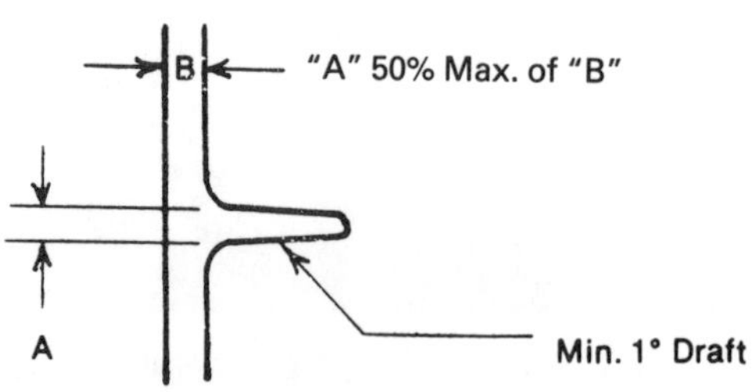

Figure 2.7 Rib-to-wall relationship

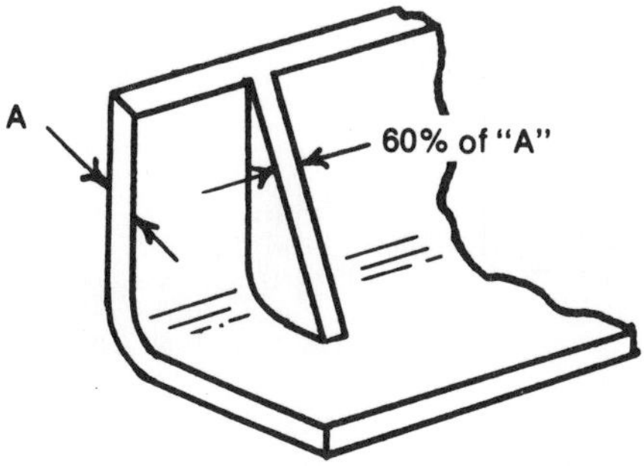

Figure 2.8 Rib wall supports

Boss height should not exceed two times the boss diameter, as shown in Figure 2.9. Bosses should always be designed with proper radii at their wall intersection and with a minimum of 1° draft.

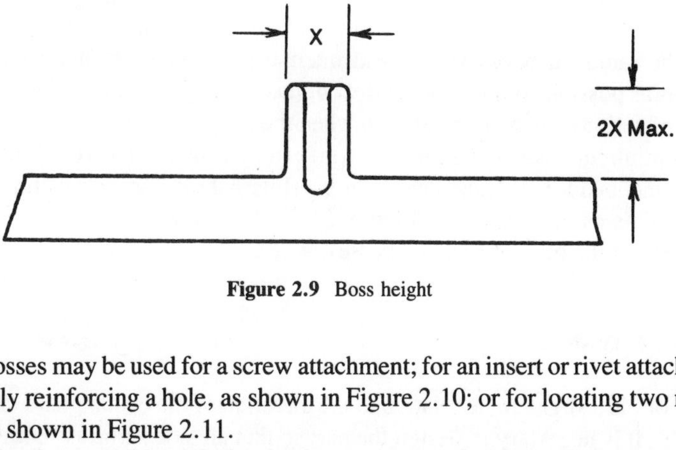

Figure 2.9 Boss height

Bosses may be used for a screw attachment; for an insert or rivet attachment; for simply reinforcing a hole, as shown in Figure 2.10; or for locating two mating parts, as shown in Figure 2.11.

It is often possible to design intricate systems of structurally interdependent bosses and ribs that give dimensional stability and provide for attachment of related components, thus maintaining proper wall thickness ratios, reducing material usage, and reducing molding cycle time.

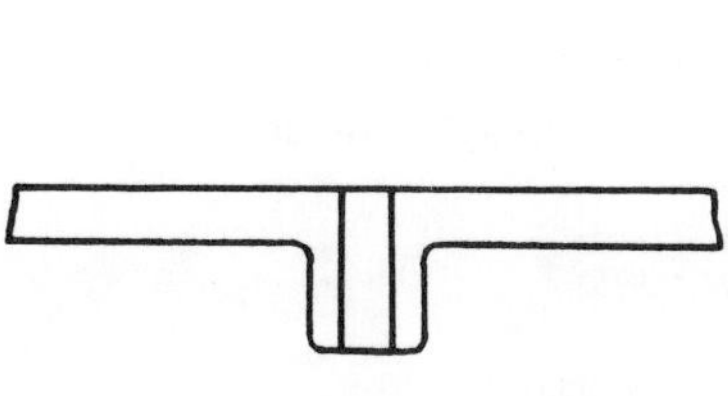

Figure 2.10 Reinforcing bosses

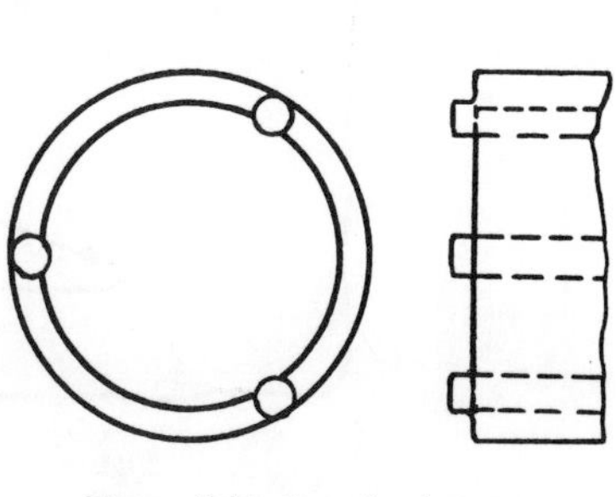

Figure 2.11 Locating bosses

Figure 2.12 Edge-stiffening methods

2.2.4 Edge Stiffening

Unsupported edges of a part may be stiffened or strengthened by turning the edge or changing the plane of the wall as shown in Figure 2.12.

These methods may be employed to provide sufficient edge rigidity and to reduce or eliminate warpage. This is especially useful when minimum wall thickness is used.

2.2.5 Radius

The radius at both external and internal plane intersections should be as generous as possible to improve the flow of the molding material and to lessen the chance of serious stress concentration when the part is under a load.

A minimum radius of 1 mm (0.04 in) external and 0.5 mm (0.02 in) internal is recommended for all plane intersections. (Internal radii are sommetimes called fillets.) This is illustrated in Figure 2.13. A minimum radius is generally considered at the base of ribs and bosses, where sink marks would be objectionable.

2.2.6 Draft

Draft is a slight angle relative to the direction of the opening and closing of the mold. It is necessary to design the part so that all side surfaces, both exterior and interior, have draft. This allows the part to be removed from the mold without sticking or rubbing, which can damage the part surface.

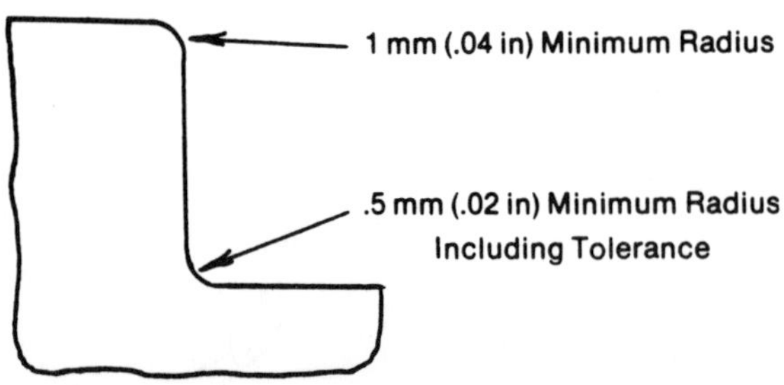

Figure 2.13 Minimum radii

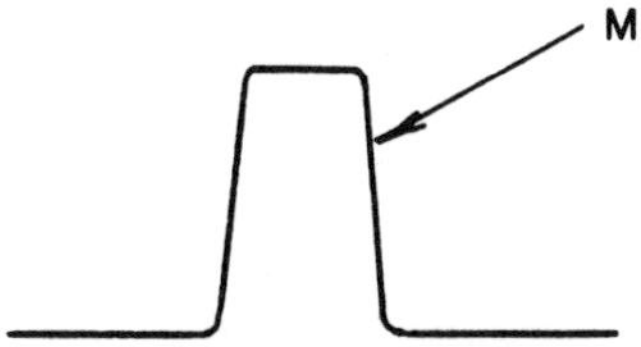

Figure 2.14 Minimum draft recommendations

A minimum draft angle of 1° is recommended on all smooth surfaces parallel to mold movement. If the mold surface is textured, the minimum draft angle should be increased 1° for each 0.03 mm of texture depth. This recommendation is for all part details such as ribs, bosses, elevation changes, and holes.

A draft angle of 1° will yield a thickness change on ribs and bosses of .017 cm/cm (in/in).

Figure 2.14 illustrates the minimum recommended draft.

A draft angle of 3° is recommended when possible. This facilitates part removal from the mold and is highly advisable, as lubricants and mold release agents are not recommended for parts to be electroplated.

If a "zero" draft condition is necessary for a part, slide or side cores in the mold must be considered that can be pulled separately from the side before the mold opens.

2.2.7 Molded Threads

Either external or internal threads may be molded in plastic parts. A rotatable insert that unscrews from the part can be used to obtain threads that are parting line-free. If a parting line across the threads is not objectionable, they can be produced without the rotating core simply by putting the threads in each half of the mold.

Molded threads should be rounded rather than sharp. Rounded threads will resist chipping and cracking and will also facilitate the flow of molding material into all areas of thread.

Molded threads should be generally larger and coarser class. A chamfer recess or counterbore as shown in Figure 2.15 should be used to avoid chipping of internal threads.

Molded threads are not recommended for diameters less than 0.25 in.

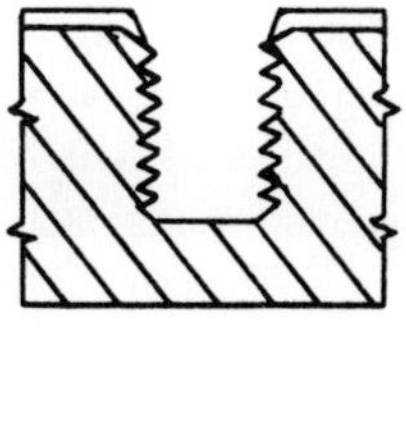

Figure 2.15 Thread chamfer

Figure 2.16 External molded thread

In all cases, consideration should be given to the effect of the electroplating deposit on the fit of the thread. Figure 2.16 shows a typical external thread with parting line.

2.2.8 Cut Threads

Thread cutting or tapping is not recommended on plastic that is to be electroplated.

2.2.9 Inserts and Fasteners

Inserts are usually metal fastening devices that are molded into the part or applied after molding. They are commonly threaded for fastening and have a knurled or grooved surface to lock into the plastic. Inserts are used when repetitive fastening and unfastening is anticipated.

It is generally not good practice to apply metal inserts to plastic ahead of plating, as the plating acids may deteriorate the insert.

Other fasteners include self-tapping screws, both thread-cutting and thread-forming, and speed nuts, clips, and rivets.

When considering threaded connections, self-tapping screws should be specified only when no more than three fastening and unfastening cycles are expected to occur. For repetitive fastening and unfastening, such as would be required for maintenance or service access at frequent intervals, inserts should be used, except when molded threads are suitable.

Figure 2.17 shows typical inserts, and Figure 2.18 shows typical self-tapping screws.

Figure 2.17 Typical inserts

2.2.10 Surface Crown

A slight crown of .010 to .015 in/in is recommended over broad surface area to reduce the field of focus as shown in Figure 2.19. This will improve the visual aspects of the finished plated part.

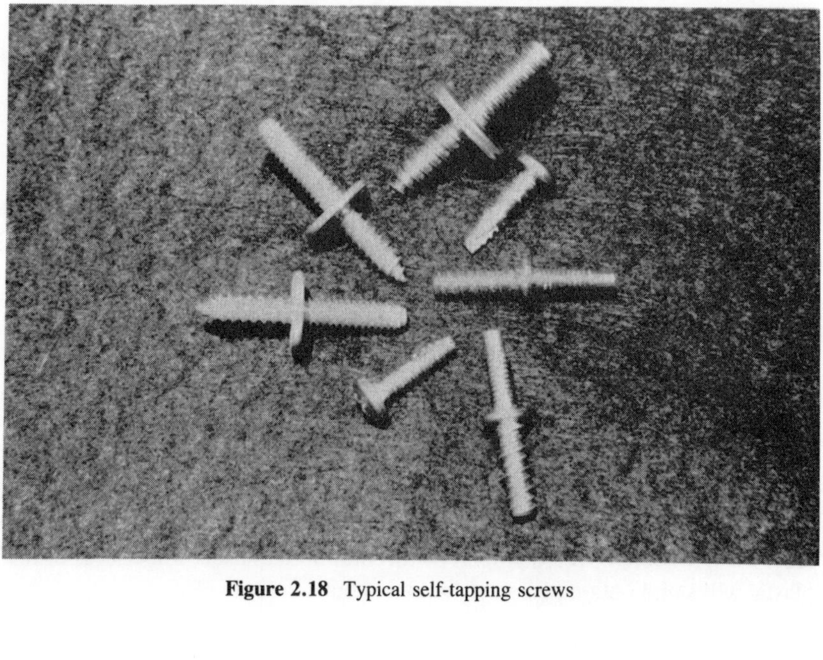

Figure 2.18 Typical self-tapping screws

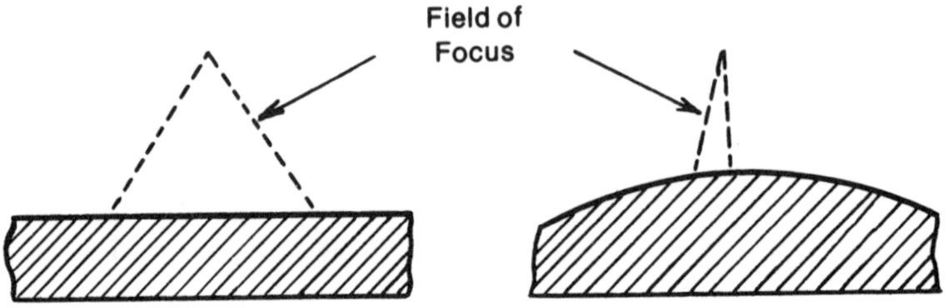

Figure 2.19 Recommended crown

2.3 APPEARANCE DESIGN

Appearance is an important factor in designing for the majority of electroplated plastic applications. In many cases the physical appearance of the plated part is an important factor in the sale of the designer's product. Bright chromium electroplating tends to make defects more visible; efforts should be made to design out appearance defects.

Gating, parting lines, and vents should be located in noncritical areas whenever possible to minimize the possibility that they will produce effects that detract from the overall appearance of the part.

Avoid mechanical or solvent welds on two-piece part designs, as both are very difficult to plate without adversely affecting the final appearance of the product.

Avoid visible sink marks by properly following recommendations on wall thickness, ribs, and bosses.

Anticipate surface appearance problems whenever possible and minimize or eliminate them by utilizing texturing or crowning techniques.

Consider the merits of selective plating to cover blemish areas, avoid corrosion problems in low-current-density areas, and avoid problems of paint adhesion on bright chromium in severe applications.

Consider the merits of decorative paint color schemes to enhance the appearance of the finished chromium-plated product. The proper use of color in decorating the part can very favorably affect the overall appearance.

2.4 ELECTROPLATING DISTRIBUTION

Proper design with electroplating distribution in mind is required to enhance the aesthetic qualities of a plated plastic part. Without the use of gently curved, convex surface, radiused angles, and a minimum of protuberances, the appearance will suffer. Sharp corners may cause plate buildup and burn on the edges. Deep recesses may not be covered properly with chromium. Surface imperfections in the plastic will stand out on planar surfaces without a crown.

2.4.1 Part Performance

Parts designed without consideration to sharp edges that build up heavily in plating may not mate properly to component parts or to assemblies. Screw and bolt assemblies may not thread properly.

Recesses in a part should always be at least twice as wide as they are deep if they are to be plated. Figure 2.20 illustrates plate thickness variations that can be obtained in recessed areas.

Since a minimum plate thickness is required on most parts, this part would have extremely heavy plate on the prominent surfaces as a result of trying to meet minimum thickness in the recess. The resulting electroplate would yield a heavier and more brittle deposit and could result in inferior thermal cycle performance. This may appear extreme, but it is one of the most common design problems encountered on electroplated parts. Consideration of current density and resulting plate thickness should always be a major factor in part design.

Corrosion performance in accelerated testing and in actual service is a direct function of the plate composite and thickness.

If the problem in Figure 2.20 is encountered in practice and a satisfactory solution cannot be found, the minimum thickness may be waived in the bottom of the recess. In this case a point halfway up the side wall of the recess may be used as the checking point for plated thickness. As a result, the bottom of the recess will be underplated and will be subject to corrosion failure. It should also be noted that not only will the plate thickness be insufficient, but the recess shape is such that moisture and corrosive solutions are trapped, furthering the potential for corrosion failure.

Proper design with regard to plate thickness distribution is also important to minimize the cost of the part. If three to four times the minimum thickness is required on the majority of the surface in order to achieve minimum thickness in the recesses, then the cost of the copper and nickel electrodeposits will be higher than otherwise necessary.

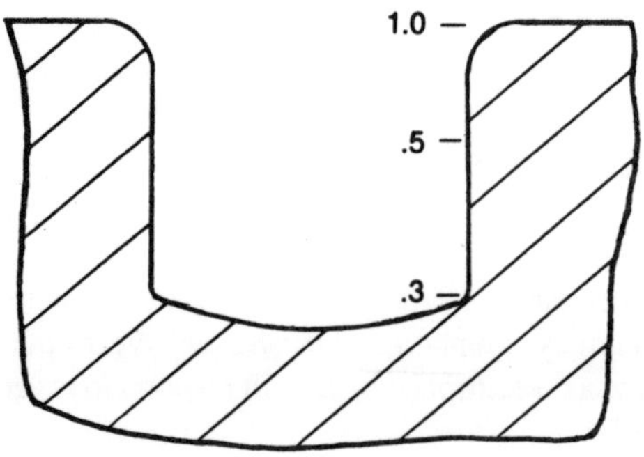

Figure 2.20 Comparative poor plate distribution ratios

2.4.2 Controlling Electroplate Distribution

The first consideration in controlling electroplate distribution is designing the part properly. The simple illustrations in Figure 2.21 show a few of the design pitfalls and how to avoid them.

Selective plating is another means of controlling electroplate distribution. This technique utilizes an acid-resistant paint applied ahead of plating to selected areas of a part to prevent plating in those areas.

The use of this stop-off (or resist) paint in the recessed or low-current-density areas of a part can completely eliminate the problems of trying to meet minimum plate thickness in those areas. This process is used extensively in producing plated plastic trim for the automotive industry.

Because of a slight fading of the stop-off paint in the plating process, it is usually top-coated with a finish coat of paint after plating.

Stop-off paint is generally applied with conventional spray techniques or by dip-coating in a fixture.

Care must be taken in the design stage to see that the line of intersection of plate and stop-off paint is located in a low-current-density area on the part. Otherwise this intersection will exhibit a nodular buildup that will detract significantly from the part's appearance. This is discussed in more detail in the section on selecting plating.

Auxiliary anodes is another method of affecting electroplate distribution. The use of these anodes mounted on the plating rack in close proximity to recessed areas raises the effective current density in the recesses, producing a thicker deposit in these areas. This technique is quite common in the industry and offers more freedom to the part designer. Auxiliary anodes, however, are not the answer to all design problems because of clearance and design complexity.

The use of auxiliary anodes generally adds to the cost of producing a part, and some manufacturers do not have the capability of using them.

Very uniform plate thickness distribution can be obtained on complex parts by using electroless (chemical) plating. This technique is, however, costly and generally applicable to engineering or functional applications where a bright decorative finish is not required.

2.5 THERMAL PERFORMANCE REQUIREMENTS

The preceding discussion of plate distribution touched briefly on the effect on thermal performance. There are other major factors, including part geometry, resin characteristics, molding techniques, preplating process, and electroplate composite, which also determine how an electroplated part will respond to severe temperature extremes.

Not This

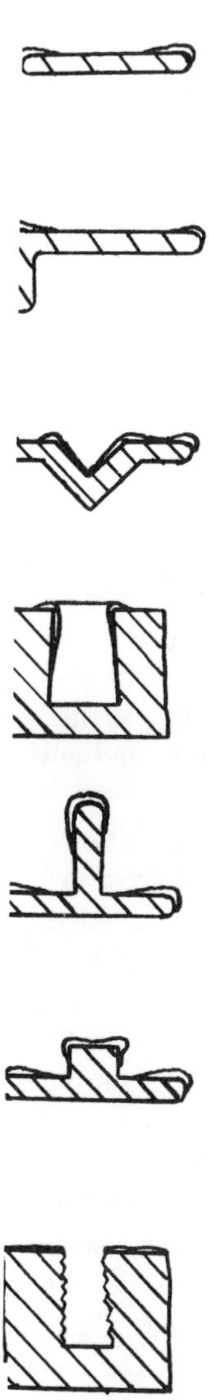

This

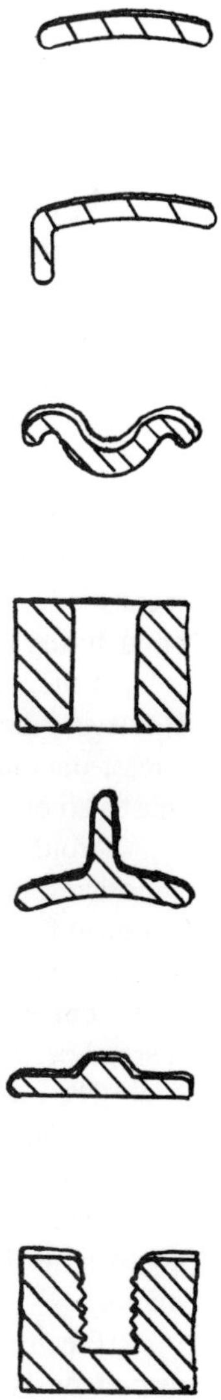

A slightly crowned surface, radiused on the end is most desirable. A .010″ to .015″ crown per inch is recommended.

Avoid sharp angles both inside and outside. Minimum recommended outside radius is .040 in. Inside radius is .020 in.

V-shaped grooves should be avoided. Provide liberal radius.

Blind holes and extremely small, deep holes will not usually accept uniform plating except in plating baths designed for high throwing power with well designed agitation of the work and bath. Depth of hole should not exceed 1 1/2 times the diameter of minor dimension.

High protuberances such as ribs and bosses should be as gradual as possible.

Lettering either raised or depressed should remain as close to the general surface as possible. As a rule, do not raise or depress from the general surface more than one-half its width.

Countersink threaded holes to minimize plateup at the aperture.

Integral studs for fasteners should be shortened as much as possible and filleted at the base. The tip should be radiused to prevent buildup.

Figure 2.21 Design pitfalls and how to avoid them

Part geometry greatly affects thermal performance. Examples of part geometry that have proven difficult in passing thermal requirements include:

Solid plastic or heavy mass-to-surface area. This can be solved by coring out the part of thinning walls.

Closed ring design. This does not allow for deflection of the plastic during thermal expansion. This becomes more critical on larger parts.

Intersection of ribs on the same plane is not recommended. Lowering one rib section as little as 1/16 in greatly reduces turbulence in molding, thus reducing stress and improving metal-to-plastic adhesion.

Resin characteristics have a major effect on thermal performance. In general, plastics have a much higher coefficient of thermal expansion than the metals used to plate the plastic. This creates great stress on the plastic-to-metal bond during thermal expansion and contraction.

Thus, dimensional stability is a prime consideration in selection of a resin. However, the other characteristics, such as moldability, plateability, strength, stress, obtainable adhesion levels, coefficient of thermal expansion, and cost, must also be considered.

Molding techniques affect thermal performance. The mold design is the first consideration in molding. Gating, runner systems, and venting all have to be designed for a favorable effect on thermal performance. The molding parameters are the next consideration. Resin must be dry. Plastic melt temperature, fill speed, mold temperature, pressure, and cycle time must all be optimized to enhance thermal performance.

The preplate system is important to thermal performance because once the resin is selected, it is also a major factor that determines the strength of the plastic-to-metal bond. Variations in the chemistry, immersion time, and temperature have varying effects on adhesion levels. Higher adhesion levels will help compensate for other problems that have an adverse effect on thermal performance.

Electroplate composite and quality also affect thermal performance. Plastic, as previously noted, has a higher coefficient of expansion than metal. The metal electrodeposit for decorative applications generally contains copper, one or more types of nickel, and chromium. When all coatings are properly applied, the bright acid copper is a very ductile deposit, the semibright nickel (if used) is semiductile, the bright nickel less ductile than the semibright, and the chromium is brittle.

The significance of this is that a ductile interface is needed between the more brittle nickel or chromium layers and the flexible, moving, plastic substrate during thermal expansion and contraction. This is the main function of the bright acid copper plate. Therefore, for good thermal performance the copper plate thickness should be 1.0 to 2.0 times the total nickel thickness. Even proper copper thickness, however, will not insure good thermal performance if plating solutions are not properly controlled or if there are major design or molding problems.

SUGGESTED READINGS

1. *Plastics Engineering Handbook*, 3rd ed. New York: Reinhold Publishing. Chapter 17, "Design of Molded Parts."
2. *Ibid.*, Chapter 18, "Standards for Tolerances of Molded Parts."
3. *Ibid.*, Chapter 17, "Design Standards for Inserts."
4. B. J. Sexton, *Part Design Parameters of Plastic Substrates for Plating*, SPE RETEC, Hartford, Connecticut, December 4 and 5, 1968.
5. *SPI Standards and Practices of Plastics Custom Molders*, SPI, 250 Park Avenue, New York, N.Y. 10017.
6. Robert L. Coombes, *Electroless Copper in Plating on Plastics*, 2nd Annual Meeting ASEP, Boca Raton, Florida, November 14, 1969.

2.6 DETAILED DESIGN RECOMMENDATIONS

It is obvious that all design problems cannot be covered with general rules and guidelines. For this reason, specific detailed recommendations are offered below in summary form with illustrations. This is intended to be a handy and fast reference source for the designer.

2.6.1 Designing Wall Thickness (Automotive Grilles)

Typical grille bar wall thickness may be varied from a minimum of 2.3 mm (0.09 in) to a maximum of 4.6 mm (0.18 in). The depth of all horizontal and vertical bars should be a minimum of 19 mm (0.75 in) and no greater than 38 mm (1.5 in). (See Figure 2.22.)

Grille bar sections having a wall thickness of greater than 4.6 mm (0.18 in) but less than 7.9 mm (0.31 in) are not feasible for plated grilles.

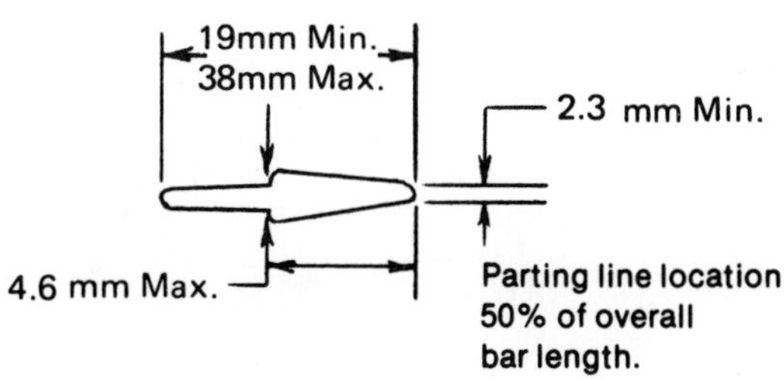

Figure 2.22

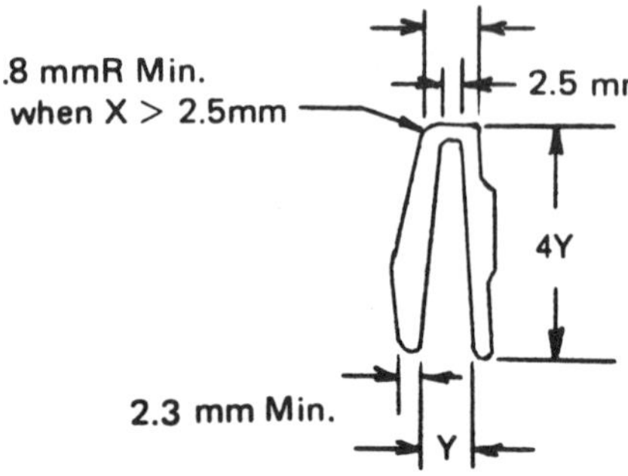

Figure 2.23

Sections greater than 7.9 mm (0.31 in) must be cored. The ratio of core base to its depth must not exceed 1 to 4 as this causes mold blade fatigue. Minimum core tip thickness should be 2.5 mm (0.1 in). (See Figure 2.23.)

The maximum cross section of an irregularly shaped bar should not exceed a 4.6mm (0.18-in) diameter circle fully inscribed within the section. (See Figure 2.24.)

2.6.2 Designing with Proper Draft

Allow 15° minimum draft when transitioning a partline line from one position to another on the same grille bar. Offset the parting line 2.5 mm (0.1 in) minimum before starting the draft. (See Figure 2.25.)

A minimum draft angle of 1.5° should be used for grille bars.

A minimum step of 0.3 mm (0.01 in) should be designed in at the parting line for mold mismatch.

Leading edges of all frontal grille bars should be crowned 0.1 mm (.005 in). If the frontal face of the grille bar measures 2.5 mm (0.1 in) or less, allow 0.5 mm (0.02 in) minimum radius for corners and edges where sharp appearance is

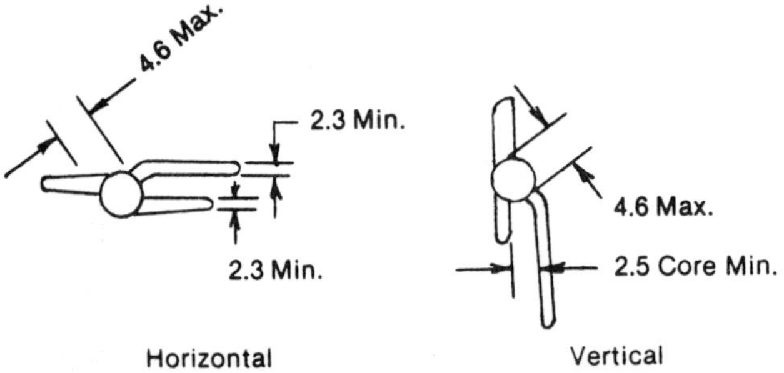

Figure 2.24

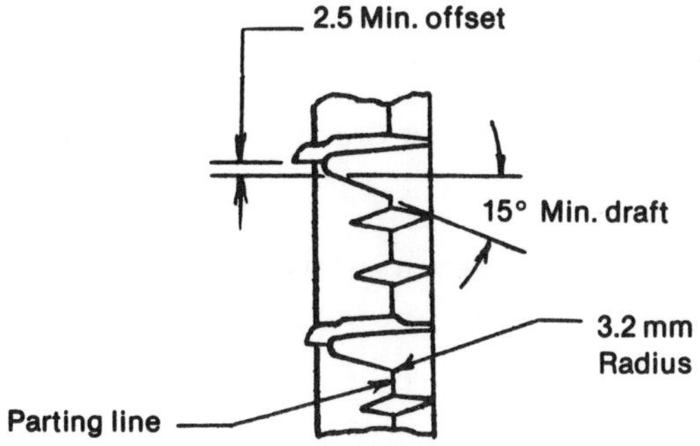

Parting Line Parameters

Figure 2.25

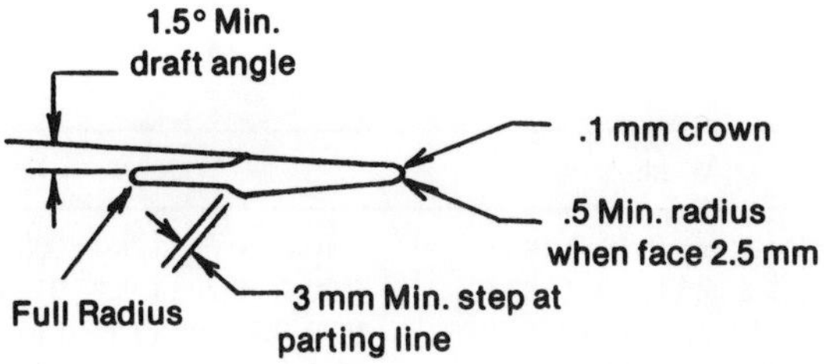

Figure 2.26

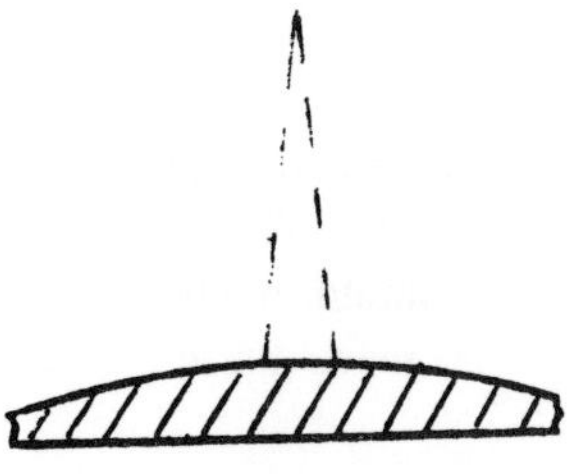

A crown is recommended over broad surface areas. The minimum recommended crown is .010 to .015 cm/cm (in/in). This reduces the field of focus enhancing the appearance of the finished part.

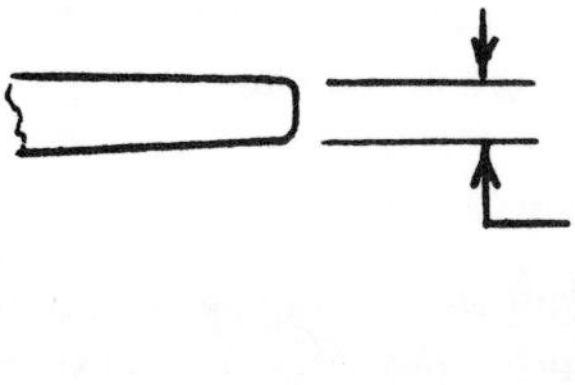

On surfaces less than 2.3 mm (.090 in) wide, such as the face of a bar in a grille, the crown may be omitted.

Figure 2.27

desired. When frontal faces exceed 2.5 mm (0.1 in) in thickness, use a minimum radius of 0.8 mm (0.03 in). (See Figure 2.26.)

2.6.3 Designing with Proper Crowning on Flat Surfaces

(See Figure 2.27.)

Recommended ratios are given in Table 2.3.

TABLE 2.3

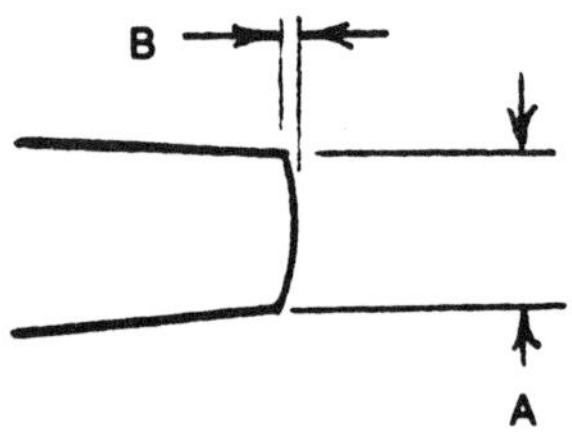

Width A	Width B
2.3–4.9 mm (0.09–0.19 in)	0.10–0.20 mm (.004–.008 in)
5.0–9.8 mm (0.20–0.38 in)	0.20–0.30 mm (.008–.012 in)
9.9–15.9 mm (0.39–0.62 in)	0.30–0.41 mm (.012–.016 in)
16.0–25.5 mm (0.63–1.00 in)	0.41–0.51 mm (.016–.020 in)
25.6–50.1 mm (1.01–2.00 in)	0.51–0.61 mm (.020–.024 in)

2.6.4 Designing with Adequate Radii

Outside-corner radii should be 1 mm (0.04 in) minimum and 2 mm (0.08 in) minimum when located at a parting line. (See Figure 2.28a.)

Inside-corner radii should be 0.5 mm (0.02 in) minimum including tolerance. A radius of 0.8 mm ± 0.3 mm (0.03 in ± 0.01 in) meets these minimum requirements and is recommended. (See Figure 2.28b.)

2.6.5 Designing to Eliminate Sink Marks

To prevent sink marks, support ribs should not exceed 60% of the material thickness at the point of intersection as shown in Figures 2.29 and 2.30.

2.6.6 Grille Attachment Design (See Figure 2.31.)

The grille-attachment locator pads should allow for drainage to minimize drag-out of plating solution during the manufacturing process. This is especially important in areas completely enclosed by grille bars, as shown in Figure 2.32.

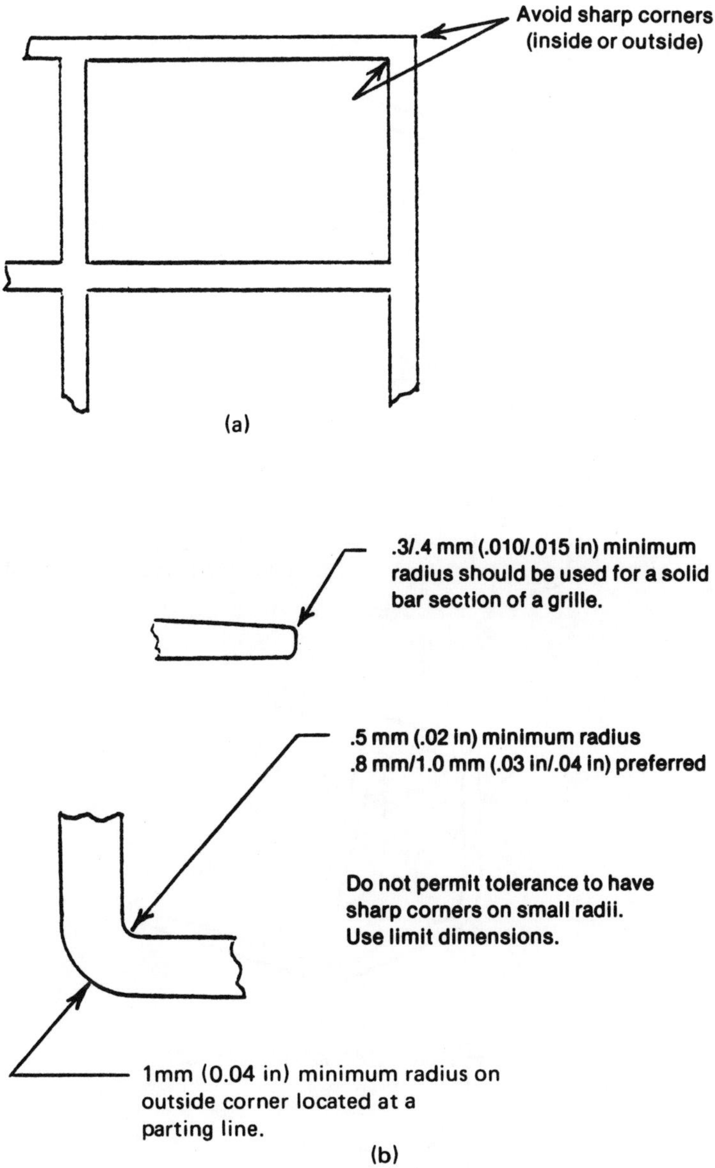
Avoid sharp corners
(inside or outside)
(a)
.3/.4 mm (.010/.015 in) minimum
radius should be used for a solid
bar section of a grille.
.5 mm (.02 in) minimum radius
.8 mm/1.0 mm (.03 in/.04 in) preferred
Do not permit tolerance to have
sharp corners on small radii.
Use limit dimensions.
1mm (0.04 in) minimum radius on
outside corner located at a
parting line.
(b)

Figure 2.28

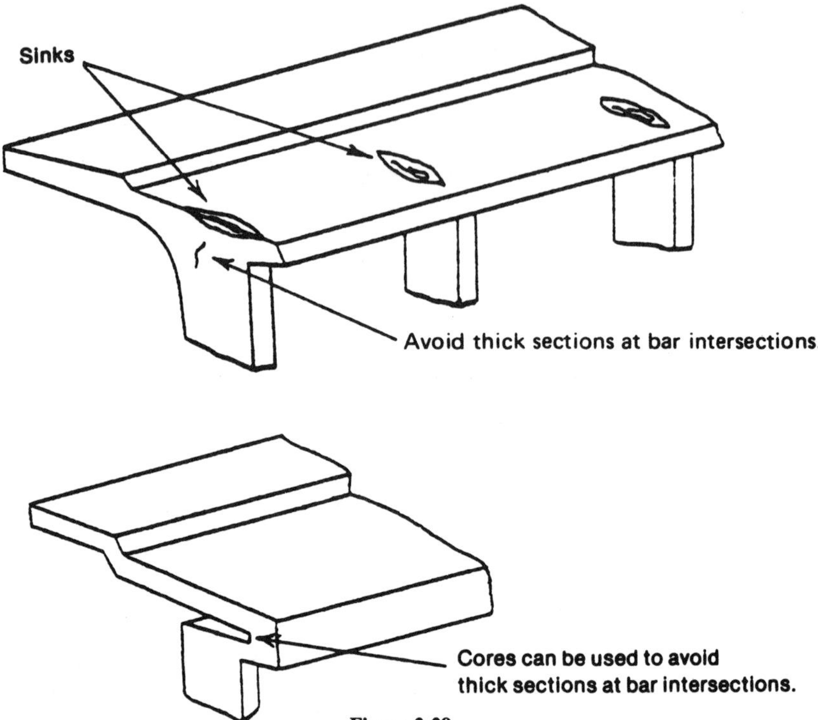

Figure 2.29

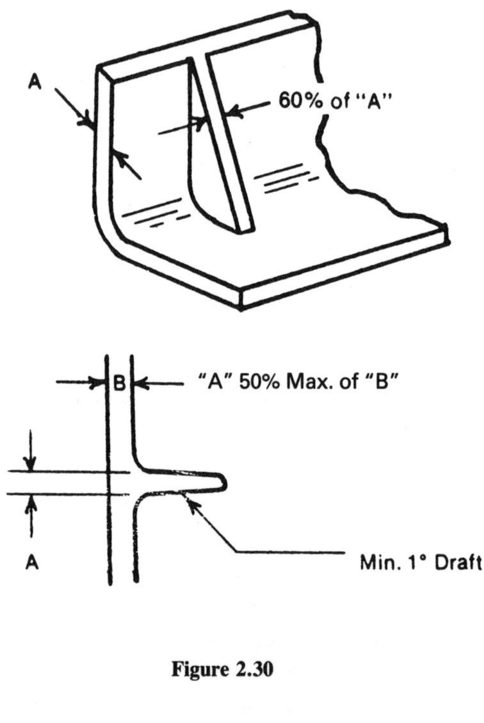

Figure 2.30

Grille mounting tabs should be supported on at least three sides as shown at right.

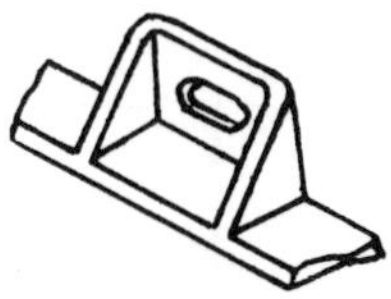

Figure 2.31

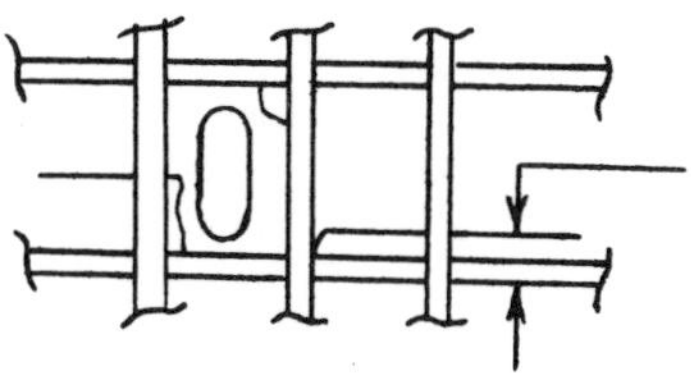

Figure 2.32

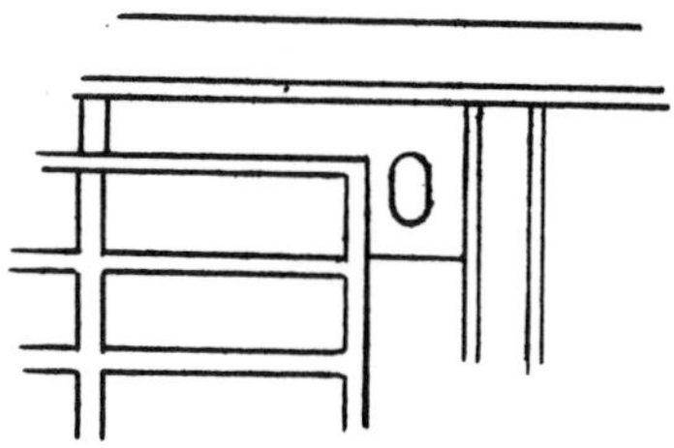

Figure 2.33

A minimum clearance of 4.3 mm (0.17 in) should be allowed around or on each side of the mounting hole or slot for screw-head clearance. (See Figure 2.33.)

Fillets of 1.5 mm (0.06 in) radius should be used in all corners of the attachment locator pads. (See Figure 2.34.)

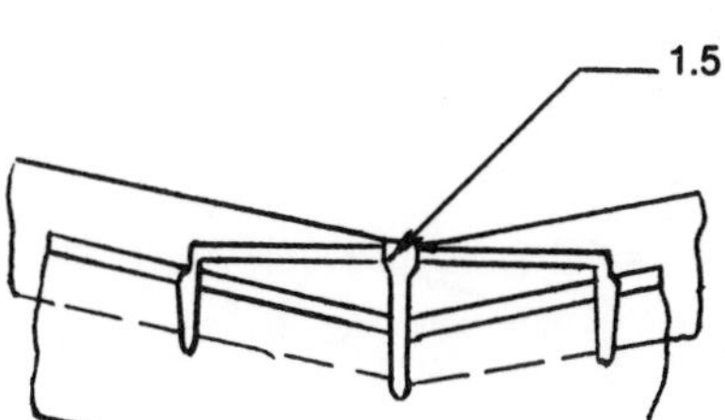

Figure 2.34

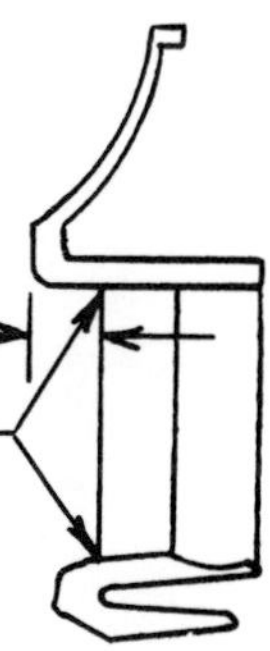

Figure 2.35

2.6.7 Designing for Plating

The depth differential between major and minor portions of the grille to be plated should not exceed 9.7 mm (0.38 in).

A minimum offset radius of 4.6 mm (0.18 in) is recommended between major vertical and horizontal bars. (See Figure 2.35.)

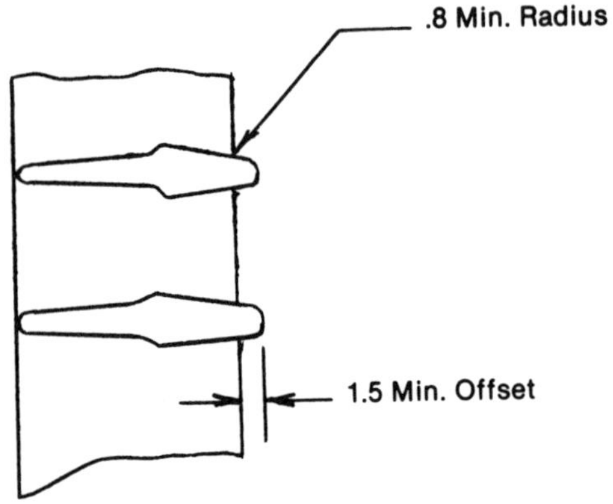

Figure 2.36

Intersections of leading edges of vertical and horizontal bars with significant chrome deposit should be offset 1.5 mm (0.06 in) minimum to prevent blistering during thermal cycling. This dimension should also include a minimum 0.08 mm (0.03-in) radius to maintain plating continuity. (See Figure 2.36.)

If plating is desired on sidewalls of typical "egg-crate" grille designs, bar spacing must be a minimum of twice the bar depth. (See Figure 2.37.)

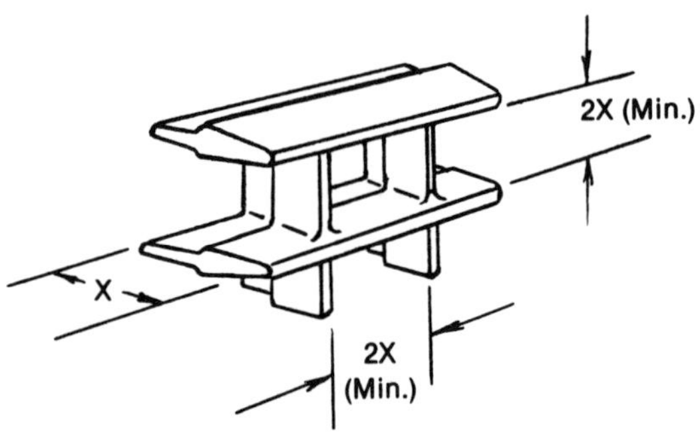

Figure 2.37

2.6.8 Designing for Paint Application

Spacing for spray-painted grille bar sidewalls should be no less than half the bar depth to achieve a minimum 45° paint spray angle from both sides of the grille, as shown in Figure 2.38.

The depth of attachment or locator pads relative to the leading grille bar and/or surrounding edges must be set so as to achieve the specified minimum coverage of bar sidewalls when painting the grille. (See Figure 2.39.)

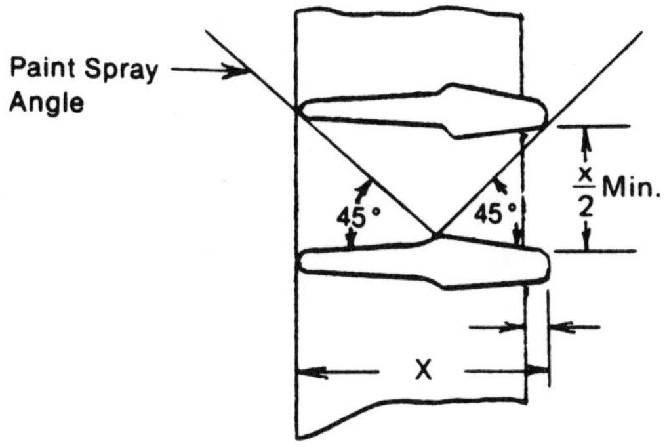

Figure 2.38

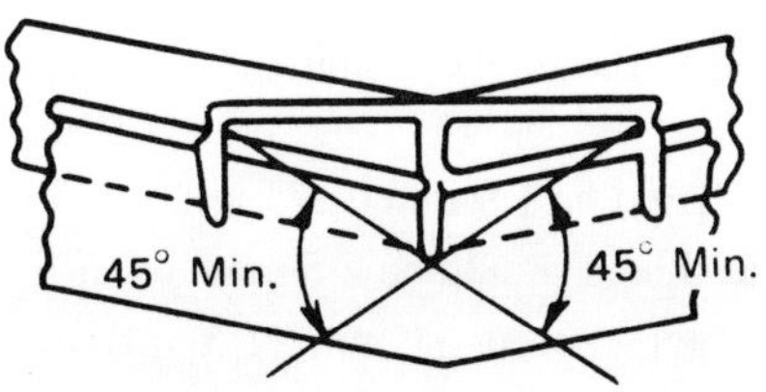

Figure 2.39 Paint spray angle

2.6.9 General Design

Plastic grilles with a vertical column theme should be designed within the limits listed here and demonstrated in Figure 2.40.

The offset between the horizontal tie bars and the vertical bars should not exceed 12.8 mm (0.5 in).

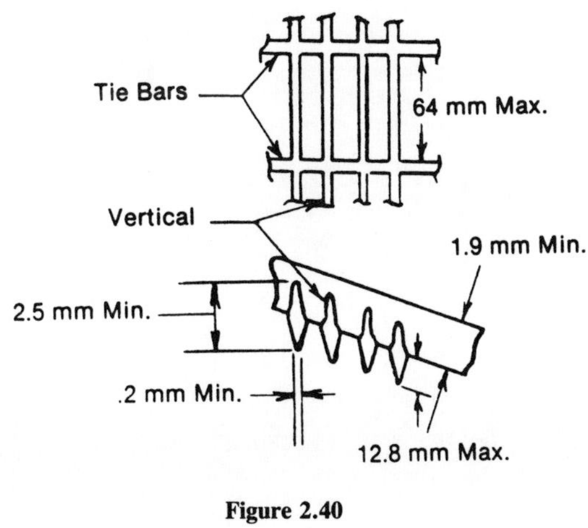

Figure 2.40

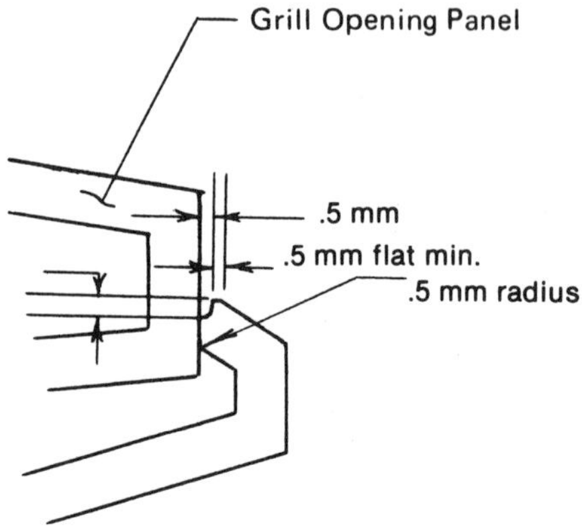

Figure 2.41

Frontal vertical bar width should be maintained at 0.2 mm (0.09 in) minimum.

Minimum tie bar depth should be 1.9 mm (0.75 in).

Minimum depth of vertical bars should be 2.5 mm (1.0 in).

Tie-bar spacing should not exceed 64 mm (2.5 in) and applies to all horizontal and vertical bars.

The minimum vertical bar spacing is dependent on decorating limitations.

Antipaint chip beads are recommended whenever the grille surround molding is styled to contact the surface of the grill opening panel. Recommendations are shown in Figure 2.41.

3

Mold Design for Injection Molding of Plastics for Plating

3.1 MOLD STEEL

The quality of the part surface appearance is directly related to the type and quality of steel used in the mold. Prehardened P20 (33-39 Rockwell) is the most popular and versatile mold steel; CSM #2 High Hardness (P-20 320-340 Brinell) is popular too. Select tool steel of the oil-hardening and H-13 type can be hardened and will accept a highly polished surface without pits. This may be selected when abrasion is expected to be a problem.

Surface finish should be specified. Use SPE-SPI mold finish comparison kit. Specify #1 finish for critical surfaces and #3 and #4 finish for core and nonappearance surfaces. All runners have to have a good polish.

3.2 SPRUE DESIGN

Sprues should be short and relatively thick—6.5 mm (1/4 in) diameter or more. Standard sprue bushings with a taper of 42 mm/m (1/2 in/ft) are satisfactory. Sprue bushing should be smooth and polished. It is a good design practice to put a cold well slug at the base of the sprue. This will trap the cooler material from the front end of the nozzle, which may cause a blemish if permitted to flow into the piece. The cold well should be slightly larger than the sprue and approximately 9.5 mm (3/8 in) deep.

3.3 RUNNERS

Runners distribute the plastic material from the sprue to the cavities. They should cause minimum cooling and resistance to flow. To do this, runners should combine maximum cross section with minimum surface area and be as short as possible. A full round runner meets these requirements better than any shape but is more difficult to machine because both sides of the mold must be cut individually and the half runner must mate when the mold is closed. The trapezoidal runner is next best. Half-round or shallow rectangular runners are not desirable because of their high surface-to-volume ratio and restricted flow area. Figure 3.1 illustrates these runner shapes.

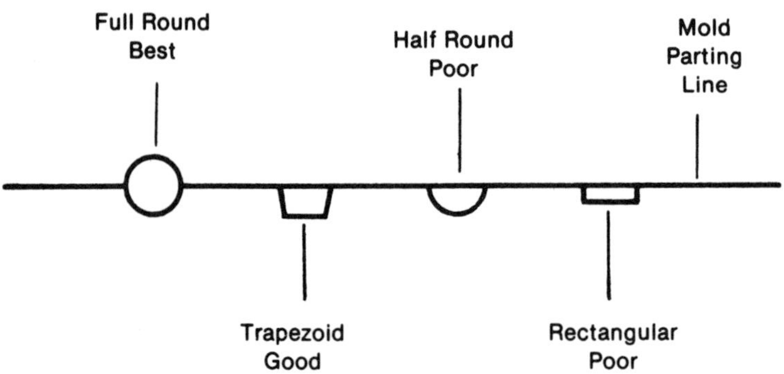

Figure 3.1 Typical runner cross sections

A general recommendation is that they should start at 9.5 mm (3⁄8 in) diameter for the primary runners and be reduced in diameter by 1.6 mm (1⁄16 in) for secondary runners. Exceptions would be runners less than 150 mm (6 in) long; for very small parts, the primary runners can start at 7.9 mm (5⁄16 in) diameter. For very large parts, such as an automotive grille, the primary runners may be increased to 4.9 mm (7⁄16 in) diameter and the secondary runners increased to 9.5 mm (3⁄8 in) diameter.

The mold should be laid out so that runners are as short as possible. Sprue and runner inserts must be matched to eliminate any rough mating lines and to insure smooth plastic flow.

3.4 GATING

Theoretically, a gate should be as small as possible, yet permit the mold to fill adequately. Gate sizes can affect the production cycle. If the gate is too large, the injection time or dwell time must be increased to assure gate freeze-off before mold discharge takes place. To determine gate sizes, start with the smallest possible size. Successive trial molding and gate changes eventually result in optimum design.

Gate size, shape, and placement affect the flow pattern of material entering the mold and may influence temperature, fill time, and overall part quality. Gates should be placed so that the material, as it enters, impinges against a wall. Gates should not be placed in such a way that they enter heavy sections or intersections of ribs.

3.5 EDGE GATING

Edge gates should have 0.8 mm (0.03 in) land with a double chisel entrance to the part cavity. The transition from the runner to the gate should be a smooth, rounded surface with no flow restrictions, as shown in Figure 3.2.

As Figure 3.3 demonstrates, the recommended gate thickness (T) should start at 1.0–1.3 mm (0.04–0.05 in) and be increased if required for fill considerations or for stress relief. There should be no reason to exceed 2.3 mm (0.09 in) gate thickness for most applications. If the 2.3mm (0.09-in) gate thickness is reached and additional gate cross-sectional area is needed to aid fill, the width of the gate can be expanded to the diameter of the runner (D), or additional gates may be added.

3.6 SUBGATING

Next to edge gating, subgating is the most popular method of gating a part. The primary advantage of subgating is that the part is automatically degated during the ejection of the part from the mold. The most stated disadvantage is the shear heating and stress created as the material is forced through the subgates.

An important consideration in designing subgates is the shear angle. The shear angle, which is the angle between the ejection surface and the subgate centerline, should be 30° to 35° for proper ejection.

Ejector pins can be placed to remove runner portions of the subgate system and will sometimes help part ejection if ejector pins cannot be adequately placed on the part for proper ejection.

Subgates should start out at 0.75–1.3 mm (0.03–0.05 in) wide and be increased for fill considerations or stress relief as required. For most applications, 2.3 mm (0.09 in) diameter is the maximum that should be considered for a subgate.

The transition from the runner into the subgate should be smooth and free of flow restrictions. This type of gate can be used where no marks are permitted on the part. (See Figure 3.4.)

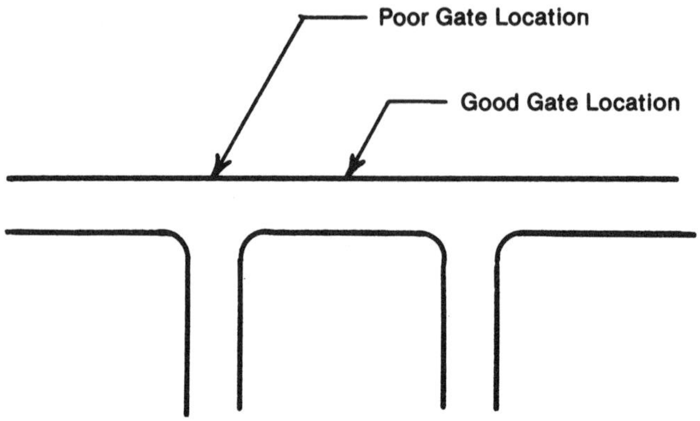

Figure 3.2

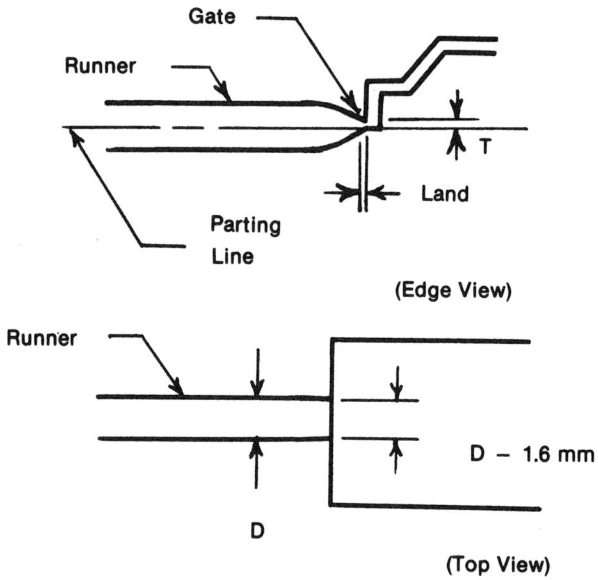

Figure 3.3

3.7 GATE AND RUNNER BALANCE

Gate and runner balance is desirable on all parts but is particularly important on larger, complex parts. To minimize molded-in stress in parts, particularly those with complex shapes, the gates and runners must be balanced so that the cavity fills uniformly and equally from all gates and runners. If the runner system is symmetrical, this balancing can generally be accomplished during the tool tryout by modifying gate sizes slightly relative to each other.

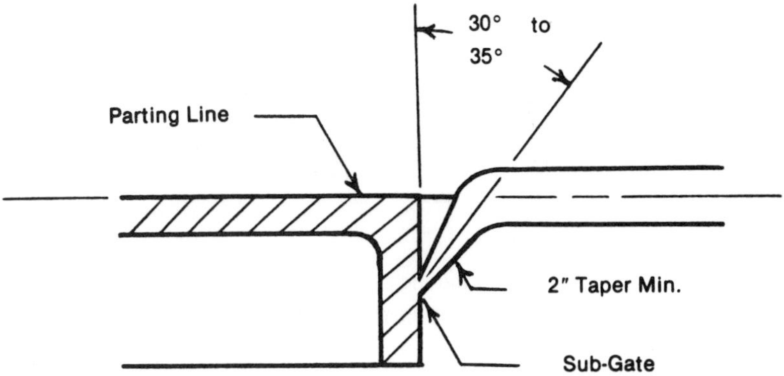

Figure 3.4

3.8 MOLD SHRINKAGE

Final part dimensions will vary material to material in the same mold dependant upon mold shrinkage. A typical mold shrinkage for different plastic materials is shown in Table 3.1 Mold shrinkage will vary from grade to grade within the same polymer, and vary dependant upon processing conditions. Raw material suppliers' literature should therefore be consulted for more specific recommendations.

TABLE 3.1

ABS	.005*
Polysulfone	.007
Polypropylene	.018
Filled Nylon	.009–.022
Modified PPO	.006

*Measurements given in cm/cm or in/in.

3.9 VENTING

Molds should be well vented as all cavities contain air which must be dispensed if the plastic is to fill the mold completely. In many molds, the normal clearance around knockout pins and nonuniformities at the parting line serve as air vents, but due to the temperature and material used in molding for plating, gases are formed that can cause voids and even burned areas on the molded part. The product of combustion at these high temperatures and pressures may cause pitting of the mold cavities.

TABLE 3.2 Maximum Vent Depth

ABS	38–50 μm (.0015–.002 in)
Polysulfone	76–102 μm (.003–.004 in)
Polypropylene	13–25 μm (.0005–.001 in)
Modified PPO	25–38 μm (.001–.0015 in)
Filled Nylon	13–25 μm (.0005–.001 in)

Vents should be placed around the parting line, inserts, and knockout pin by milling slots. The depth will depend on what material is being used as shown in Table 3.2.

Vents should be 25 mm (1 in) wide, if possible, and in a nonappearance area of the part.

The general recommendation is to vent the tool initially at the predicted knit line (last area of the cavity to fill) and to add additional vents in other areas only as required. Vents as close as 60 mm (2.5 in) around the periphery of a part are common.

Vents should be designed to allow gases to escape to atmosphere as shown in Figure 3.5.

Figure 3.5

3.10 WATER COOLING

Molds are cored for the circulation of water to provide adequate control of mold temperature. Good mold temperature control is important for uniform cooling of the part to minimize stresses in the part.

Size and location of water lines should be carefully considered. For instance, water lines that are too close to the surface can result in hot and cold spots due to low coefficient of heat transfer. One-half inch diameter lines spaced equally with separate inlets and outlets for each cavity and core are a requirement. Zone heating and cooling of large areas have been successfully used to obtain even flow and reduce molded-in strains. When planning a mold of this type, the lines are drilled vertically with no cross connections. (*The sprue area is normally maintained at lower temperature than the outside of the part so the material will remain warm.*) This will allow the part to be filled with reduced injection pressure. However, cycle times become a consideration if this area is kept too warm. Mold temperature can affect the adhesion on the plated part. Therefore, if adhesion is critical on a given part, the mold temperature should be optimized by trial and error, using adhesion or thermal cycle tests as the test criteria.

4

Molding Considerations

4.1 MOLDING

Molding for plating differs from "mold only" in several important ways. Problems common to all types of plastic to be plated are:

4.1.1 Surface Appearance Considerations

As mentioned in Chapter 2 (Part Design Considerations), the surface of a molding which will be plated in a decorative finish (i.e. bright chromium), must be free of essentially any visual defects. Weld lines, sinks, dull poorly packed areas, flash, gas burns all are likely to show when plated [with] and in most cases highlighted by the bright reflective plate. Plating with satin nickel or velour nickel and other non-bright finishes and the duller electroless-only plates such as EMI-RFI shielding do not require the same level of defect-free surfaces in molding.

4.1.2 Surface Composition Considerations

It should be recognized that the plating of plastic involves treatment of the surface and the immediate subsurface. Most plating of plastic requires that the surface be treated in such a way chemically, so that a porous surface is created, and that the metal will "plate into" the surface, creating a mechanical bond.

In order for this to be possible the plastic surface must represent a uniform composition of the two or more components which are present in a plating grade plastic. In the case of ABS, the primary two components are acrylonitrile-styrene and butadiene-styrene. In ABS these two co-polymers are uniformly mixed and cross-linked in such a way that the molded surface always contains a fairly good mix. This cross-linking and molecular polymer level of alloying is what makes ABS the most easily molded and plated plastic. However even with ABS there are pigments, flow agents, antioxidents and other ingredients which, when the plastic is subjected to high temperatures and high sheer rates in the molding operation, may cause a separation of components in such a way that the finished molded surface is not uniformly constituted. This often results in "hard to etch" edges at the last to fill areas in mold. Venting of the mold is very important (see Mold Design chapter 3 and Pre-plating chapter 7, as well as the importance of a slow fill speed in this chapter).

When molding most other platable plastics, this requirement of providing a uniform composition throughout the surface of the part becomes much more challenging, this is because at present there are few if any other platable plastics that are constituted and thus mold like ABS. The ABS alloys perhaps come closest; however, as the alloying component (polycarbonate, polysulfone, polyphenylene oxides, etc.) becomes a higher percentage of the mix, the molding becomes more difficult and surface uniformity suffers.

There is a family of platable plastics where the basic polymer is combined with a "filler," typically a mineral. Mineral reinforced nylon is an example. Glass filled materials are also common. These "filled" materials are much more difficult to mold with a uniform composition surface. With careful part design, mold design and molding control they can, however, be very successfully molded and plated.

4.1.3 Internal Molding Conditions

While the surface condition of a molding for plating is very important, the "inside" is perhaps equally important. Most parts must perform some function other than "just look good." They must mate with other parts and therefore must be molded (and plated) to a given size within certain tolerances. When molds are built, they often don't come out exactly right in matching the shrink rate of the molding to the desired finished part size. Therefore the molding is often "adjusted" to make the part fit either the mating part or the print. This can cause problems in plating. When it is necessary to increase the size of a molded part it will require overpacking, often with a fast fill and a relatively cold barrel temperature. This overpacking forces stress into the part and two things happen in and after plating. First, a cold melt and fast fill will create a hard glossy surface which will result in lower plate adhesion, as much as 1⁄10 that achievable with good molding for plating practices. Second, the molded in stress will begin to relax when the part "sees" the hot plating solutions. This relaxing causes movement which may cause blisters or loss of adhesion in the plating process, or if not in the plating process, almost certainly in the thermal-cycle testing. While molding parts to a smaller size may not, depending on the part configuration, mold in stress, it will likely result in sinks or a non-fill condition.

4.2 RESIN SELECTION

Care must be exercised in interpreting the physical properties listed for unplated resins, as they often change when electroplated. Impact resistance, for instance, is often greatly reduced.

TABLE 4.1 Resin Categories

General Purpose	Engineering Plastics High Strength, High Temp., High Impact
ABS	Filled Polysulfone
ABS Alloys	Filled Polycarbonate
Modified Polyphenelyne Oxide	Filled Nylon
Polyetherimide	Filled Polyester
	Filled Polyphthalamide

Flame retardant grades, while some are platable, always lower the adhesion over non retardant grades and also often cause difficulty in molding from film buildup in the cavities.

4.3 EQUIPMENT SELECTION

General-purpose ABS resins are the simplest to mold, preplate, and electroplate; consequently, they are in widest use. The other resins often are more difficult to mold for plating and sometimes require molding equipment with special features such as vented barrels, program injection, and high temperature mold heaters. Reciprocating screw machines in good repair are recommended. It has been found that screw machine moldings give about a 60% improvement in metal-to-plastic adhesion over parts molded in a ram machine when optimum etch conditions are used for each type of molding. The difference in adhesion performance can be as great as 170% when etching solutions and conditions developed for screw injection moldings are used for parts molded in a ram machine.

4.3.1 Drying

It is of utmost importance when molding for plating that the resin be thoroughly dried prior to molding. Upon heating, moisture absorbed in the resin is converted to steam. If not driven off prior to injection, this steam is compressed and is forced into the part expanding into bubbles which, at the surface, form blisters or teardrop-shaped splay marks. These splay marks often appear immediately after electroplating and may be misinterpreted to be plating blisters. They also appear at times only after thermal cycle testing. Dehumidifying hopper dryers are recommended. Hopper dryers are available that have sensing devices which insure that hoppers are maintained full, allowing enough residence time in the hopper to insure adequate drying. Moisture alarms are also recommended to be used with hopper dryers.

4.3.2 Mold Temperature-Control Devices

In maintaining consistent molding parameters, it is essential to control with precision the temperature of the mold. Therefore, adequately sized and controlled mold heaters and coolers are a prerequisite for quality molding for plating.

4.4 PROCESS CONTROLS

The ability to mold with consistency demands control of the molding parameters of temperature, pressure, and cycle time. Parameters recommended by resin suppliers should be followed as a starting point.

4.4.1 Temperature Control

All heating zones on the machine should be controlled by pyrometers and the indicated temperature recorded at regular intervals. Temperatures should be controlled on both sides of the mold by mold temperature-control units with indicating pyrometers.

Plastic melt temperature has a great influence on the metal-to-plastic adhesion in the finished, plated part. In general, the melt temperature must be sufficiently high to fill the mold easily under slow injection speed conditions to produce a stress-free part.

Low melt temperatures demanding high injection pressures and high fill speeds are to be avoided. The material supplier recommendations for optimum melt temperatures in molding each type of plating-grade material should be consulted. In general, the higher end of the range should be used while exercising care not to degrade the resin.

4.4.2 Pressure Control

All pressure settings, such as clamp, injection, hold, and ejection, must be controlled. Particular attention should be paid to the fill time (which is controlled by the injection pressure), as this is the most important single constant in molding for plating. Any changes in other settings (such as melt temperature and mold temperatures) will also be reflected in the fill time. (The fill time is the time measured by stopwatch, from when the plunger or screw starts to move forward until it stops.)

4.4.3 Fill Speed

Among the most important molding variables for the production of high-quality parts for electroplating is the injection (fill) speed. Fast fill rates will generally give parts with a high level of stress. Those parts, when plated, will not have satisfactory adhesion and will not meet thermal cycling specifications. For best results, the injection speed should be as slow as possible in order to produce moldings with the lowest possible molded-in stresses. For effective injection speed control, small shot capacity molds should not be used with large machines. It is desirable that the total shot weight (including sprues and runners) should not be much less than 50% of the machine capacity, depending on the resin's tolerance to heat history.

The optimum fill speed will depend on the part size and stock temperature being used. In general, the average required fill time should be at least double the time necessary to fill the same mold for parts not to be plated. Any additional reduction in injection speed would be beneficial in assuring maximum adhesion.

4.4.4 Process Controllers

Process-control devices that control with precision all the preceding variables are highly recommended.

4.4.5 Record Keeping

It is recommended that a record system be set up as follows:

1. Record the molding machine number, date, shift, and time the parts are molded. The data should be marked on the parts and/or their containers.

2. All important molding variables, such as stock temperature, fill time, cycle time, ram pressure, plastic lot control numbers, and so on, should be recorded in a log at regular time intervals and when any change is made.

3. Record resin lot numbers on the molding log, noting the time when a new lot is fed to the machine.

4.4.6 Incoming Resin Inspection

The physical properties of resin vary from lot to lot. It is recommended that each incoming resin batch be sampled either by laboratory means, such as measuring the melt index using an extrusion plastometer (ASTM D-1238), or by trial molding under controlled conditions with an appropriate test mold such as a spiral flow mold.

4.4.7 Contamination

Foreign resins are the most frequent contaminants in the molding operation. Compounds previously used in the molding machine may be foreign to the plating resin and constitute serious contaminants. There are differences in the conditioning cycle for preparing various plateable resins in the electroless plating. Thus, even a plateable grade of nylon is a serious contaminate in the ABS, and vice versa.

Purging foreign resins, particularly resins that have a higher melt temperature, is very difficult. They will often remain in the machine after hundreds of shots have been made. This is why test shots with plating-grade material run on machines that have not been molding the same or similar material are so often failures. The only sure way to avoid this problem is to disassemble the machine and clean the injection system manually. When a machine is used only for molding resins in the same family, such as colored ABS and plating-grade ABS, purging may be accomplished satisfactorily without disassembly, provided a foreign purging compound such as an acrylic is not used.

4.4.8 Handling Molded Plastic for Plating

Clean and careful handling of molded plastic parts is imperative, since anything less may result in inferior plated parts. The slightest blemish on a molded part may be greatly magnified by the electroplated finish. While not mandatory, personnel should wear cotton gloves in removing the parts from the mold and for subsequent handling. Fingerprints and machine lubricants leave a film that may cause skip plating, particularly if the plating system to be used does not include a cleaning bath as its first step. Parts can also be contaminated by grease and oil in and around the molding equipment, particularly in molds that have moving parts.

Mold-release agents should not be used when molding parts for electroplating. Parts for plating should be kept out of any area where mold lubrication is under way. Mold-release agents, especially silicone sprays, can contaminate the air and settle on parts.

Degating of parts requires particular attention. Caution should be taken when removing runners not to leave surface defects. Occasionally, runners may be left intact to provide a means of racking during the plating process.

Plastic parts removed from the mold should be adequately protected in shipping containers to prevent scuffing and scratching. Tissue paper, cellophane, or polyethylene bags are satisfactory for individual wrapping. (Where polyethylene is used, parts should be cooled to room temperature before packaging to prevent surface contamination.)

Finished parts are usually wrapped individually or in compartmented containers and shipped in strong packing units to the plater to minimize defects. Many molders who are also platers are using the semibulk method of preplating, which enables them to place parts on preplate trays at the molding machine.

4.4.9 Visual Inspection

A visual inspection of a molded part is by far the most meaningful test to which a part may be subjected. For the most part, the plating of plastic is a surface phenomenon. Therefore, with few exceptions, appearance defects in the finished part that are caused by the plastic substrate can be seen in the surface of the plastic prior to plating. Any defect in the plastic substrate surface will be highlighted by plating.

Typical molding defects that will show up as greater defects after bright chrome plating are weld lines, scratches, sinks, gate blushes, and flash and splay marks. Many of these defects are traceable to poor part or mold design.

4.4.10 Glacial Acetic Acid Test for ABS

The Glacial Acetic Acid Test is a useful tool that enables the molder to evaluate stresses and surface uniformity in the molded part.

Cracking from the acetic acid test is an indication of high stress levels. Cracking will often converge on gate areas and through other stress areas such as sharp corners and abrupt changes in thickness. The visible severity of the crack pattern can be used as a guide by the molder to enable the molder to improve the molding. A highly stressed part is more likely to warp when exposed to the hot solutions used in plating. It will also be more likely to fail in a thermal cycle test because the molded-in stresses will relax, causing a greater movement of the plastic, which in turn puts a greater stress on the metal-to-plastic bond.

The Glacial Acetic Acid Test will show whether the rubber phase of the ABS is uniformly dispersed over the surface. Some surface whitening may occur uniformly on an ideal molding. Poor moldings will have heavy white areas and dark shiny areas on the same part. Some preplate systems are unable to etch the shiny areas without overetching the white areas, which results in poor adhesion in one area or the other. Refer to Chapter 10 for the Glacial Acetic Acid Test procedure.

4.4.11 Ethyl Acetate Test for Polysulfone

The Ethyl Acetate Test is used for the same purpose as the Acetic Acid Test but is applicable to polysulfone rather than ABS. Refer to Chapter 10 for the procedure.

4.4.12 Methyl Ethyl Ketone/Acetone Test for ABS and Modified PPO

The Methyl Ethyl Ketone/Acetone Test is used for the same purpose as the Acetic Acid Test and is applicable to either ABS or modified polyphenylene oxide. Refer to Chapter 10 for the procedure.

4.4.13 Dye Absorption Test for Mineral-Reinforced Nylon

The Dye Absorption Test is a useful tool that enables the molder to judge the surface absorption of a molded nylon part. The rate of absorption relates, to some degree, to how the surface will respond to etching and catalyst absorption. Refer to Chapter 10 for the procedure.

5

Selective Plating Considerations

5.1 SELECTIVE PLATING

Selective plating is a specialized method used in plating plastic parts that utilizes a resist or stop-off paint to prevent plating on certain areas of the parts. This paint, when sprayed or dip-coated onto parts, can be processed through a plating system without losing its integrity and without being electroplated. Figures 5.1 and 5.2 illustrate the steps of this technique with close-ups of an automotive grille.

Figure 5.1 Resist-coated grille

Figure 5.2 Finished grille

Selectively plating plastics can provide advantages in appearance, performance, weight, ease of manufacturing, and cost and provides more freedom in design.

5.2 APPEARANCE

Customer acceptance of chromium-plated trim is well established. The consumer has demonstrated a preference for the durable, reflective qualities of chromium by a repeated willingness to pay a premium for this finish. Selective plating is one way of more easily and economically achieving this finish on a greater variety of plastic parts. Application of a contrasting finish in the proper areas can further enhance the plated appearance of highlighting the chromium.

5.3 PERFORMANCE

The proper use of selective plating can, in some cases, improve the properties of a finished part or achieve properties otherwise unattainable.

5.3.1 Impact

Impact resistance of an automotive grille, for example, is improved by resist-painting the rear portion of the horizontal and vertical bars. Work by J.P. Saxe presented at the Fifth Annual ASEP meeting[1] verifies the impact advantage of having resist paint on the opposite side of impact. Saxe concludes from this work that crack propagation begins on the side opposite impact and that there is a definite impact advantage when this side is resist-painted rather than plated.

When a fully plated grille is impacted from the front, a crack in the plate on the rear of the bar would cause stress concentration at one point in the plastic. This results in failure at low levels of impact. By resist-painting the rear of the bar, the stress from the impact is distributed over more surface area, resulting in higher impact resistance. J.J. Martin, in his work presented to AES in 1972,[2] also confirms the impact advantage of resist-painting the back half of both ABS and polypropylene automotive grilles.

Severe impact requirements on flat surfaces can be met in many cases by properly designing reinforcing ribs on the back and resist-painting the back surface. These are important considerations to the automotive designer.

[1]J.P. Saxe, "Substrate Contribution to Impact Performance of Plated Parts," Fifth Annual ASEP Meeting, November 11, 1972.

[2]J.J. Martin, "The Selective Plating of Plastics," *Plating*, September, 1971, p. 888.

5.3.2 Corrosion

Corrosion problems can sometimes be avoided by using selective plating. Many complex parts have areas that are recessed or otherwise difficult to plate with adequate plate thickness. Attempts to plate these parts completely can lead to premature corrosion failure when the part is in service. The proper use of selective plating can avoid plating in these areas, thus assuring good performance of the part from a corrosion standpoint. This has been a major consideration in specifying selective plating for exterior automobile trim such as grilles.

5.3.3 Thermal Cycle

Potential thermal cycle problems may be designed around by utilizing selective plating. Some parts may require extremely heavy sections for strength requirements. Such heavy sections are generally not conducive to good thermal cycle performance. Resist-painting these areas can minimize the problem. Care must be exercised in designing large parts with a large area that is resist-coated when this area is surrounded by an uninterrupted framing of electroplating. The effect of thermal expansion and contraction on such parts is difficult to predict.

5.3.4 Electrical Shock

Electrical shock hazards can be avoided by the proper use of selective plating. Selectively plating the inside of hand-tool and mixer housings is an example. This technique is used by the appliance industry on TV knobs, where a decorative plated knob is desired but where the shank of the knob has to be nonconductive to avoid shock hazards.

5.3.5 Paint Adhesion

Performance of decorative-finish paint is enhanced in some cases by selective plating. The adhesion of the finish paint to the resist paint is better than that of the finish paint to the chromium. This is a design consideration for parts that are to see extremely severe service. The primary reason for using selective plating in many automotive applications has been to insure good adhesion of decorative paint coatings on these parts.

5.3.6 Weight

Reduced weight is another desirable property of selectively plated plastic. The weight of plated plastic in general is only one-sixth that of plated zinc die castings. Selective plating further reduces that weight by eliminating areas of copper, nickel, and chromium deposits. Considering normal exterior plating thickness, the weight savings by selective plating is approximately 305 g/m^2 (1 oz/ft^2) of area resist-painted.

5.3.7 Manufacturing

Ease of manufacturing is a plus for selectively plated plastic. Plating rack design is greatly simplified in most cases because of the reduced area for plating on the part. Less area means less electrical contacts on the rack for carrying the current to the part. Fewer contacts means faster racking and unracking, which speeds up the operation. Fewer contacts also means lower tooling costs.

In many cases, more uniform thickness of the electroplate is achieved by resist-painting the proper areas on a part. This technique is used to prevent one area of the part from robbing current from an adjacent area, thus allowing more uniform plate deposit. In many cases, exotic and expensive auxiliary anodes are minimized or completely eliminated.

Resist-coating of hollow bosses helps prevent splitting the bosses when driving studs or screws after plating.

Resist-painting mounting pins or tabs makes heat staking easier and more uniform, insuring firm attachment to the mating part. Plated pins or tabs sometimes break or otherwise distort before long enough heat can be applied to stake the parts together properly.

Selective plating also makes possible the attachment of plated plastic bezels to acrylic lenses by ultrasonic welding, giving a reliable bond. Acrylic cannot be ultrasonically welded to a fully plated bezel.

5.4 COST

The cost of selective-plating techniques vary, depending on the area to be coated with resist. There may be a cost penalty in some cases to achieve performance characteristics otherwise unattainable. More commonly, savings are achieved due to the elimination of large areas of electroplate. The additional costs to apply resist paint, in many cases, are offset by savings in copper and nickel that would otherwise be plated on the part. Further savings may be realized in less costly plating racks, increased racking speed, and better plate distribution.

5.5 DESIGN FREEDOM

The design engineer utilizing selective plating has a new dimension of freedom in design. Extremely large lightweight parts can be easily manufactured with selective-plating techniques. Extremely complex shapes and auxiliary parts can be molded as one piece and selectively decorated with bright chromium and contrasting finish paint.

Electrical circuits could be plated on the back side of dash panels or instrument panels at the same time as the front of the part is decoratively plated.

5.6 DESIGN GUIDELINES

Now that the merits of selective plating have been reviewed, the design requirements and guidelines should be mentioned. If a part is to be selectively plated, provision should be made for this in the design of the part to insure good visual quality.

One problem that can arise with selective plating is unique to this method of decoration. If the boundary of the plated area lies on a flat surface, the boundary of the plated metal is likely to be visibly ragged and rough.

This is caused by a higher current density at this point than other areas of the part during the plating operation. (See Figure 5.3.) As it is plated, the metal tends to encroach on the resist area of the part, producing a rough edge. If the boundary of a plated area coincides with the edge of the part or occurs at a groove or undercut provided for this purpose, the current density at this point is lowered and a boundary is obtained that is sharp and free of irregularities.

In designing parts for selective chromium plating, avoid stopping off the plating on a flat surface. If this cannot be done by taking advantage of the natural configuration of the part, then the boundary of the plated area should be defined by masking grooves or steps.

The simple groove design shown in Figure 5.4 serves to avoid nodular plate at the boundary and also makes the fit of the paint mask less critical. In a design of this type the width is usually equal to the depth and normally should be 1.5 mm (0.06 in) minimum.

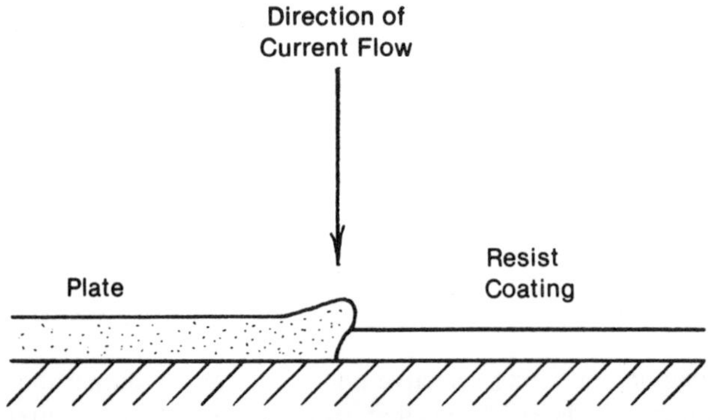

Figure 5.3

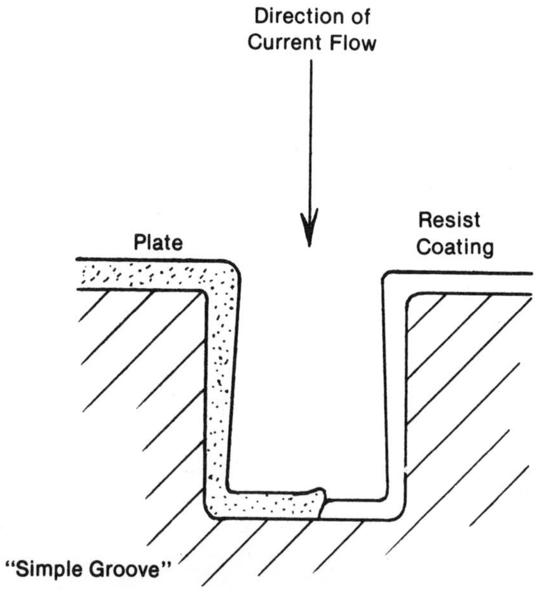

Figure 5.4

Another option is the step groove technique shown in Figure 5.5, which can be used instead of the simple groove.

Both of these designs are appropriate for working a boundary into a flat surface. Each of the two steps in this case should be 0.75 mm (0.03 in) deep minimum, and the bottom of the groove should be 1.5–2.0 mm (0.06–0.08 in) wide. The resist-paint mask fits the bottom of the second step so that the resist boundary occurs in the bottom of the first, as shown in Figure 5.6. The part is painted with a finish coat of decorative paint that covers the bottom of the groove. This completely hides the boundary area.

Another variation of the groove design is the V groove shown in Figure 5.7. This design is useful where there is a change of plane in the surface and where the step or the simple groove design is objectionable to the designer. In this case the lines are softer and not as abrupt. The design still provides a slightly recessed area for the boundary. The groove in this case is a 45°-angle groove with a minimum depth of 0.75 mm (0.03 in). The resist-paint mask is designed so that the plating boundary is in the bottom of the groove. When the part is finish-painted, the boundary line of the plating is overlapped slightly, as shown in Figure 5.8, so that it is not visible.

Another design that is used on automotive grilles is the simple step. This design utilizes a 0.75 mm (0.03-in) step for resist masking as shown in Figure 5.9. For close grille work this step is recessed 6.4 mm (0.25 in) from the leading edges

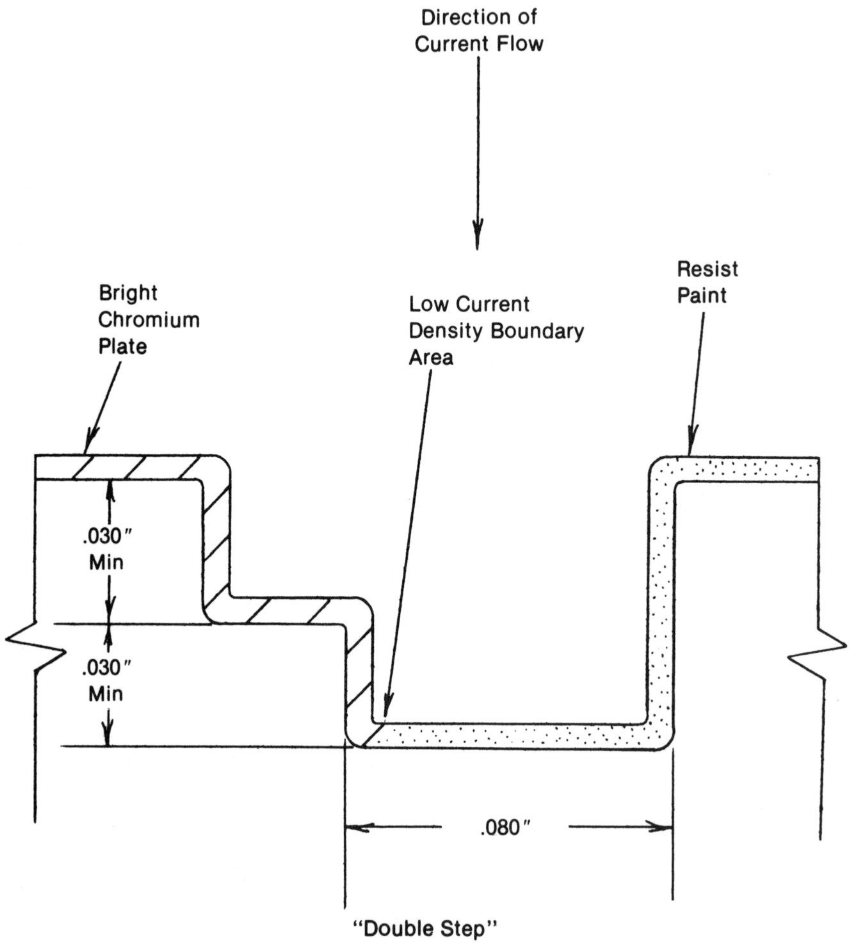
Direction of
Current Flow
Bright
Chromium
Plate
Low Current
Density Boundary
Area
Resist
Paint
.030″
Min
.030″
Min
.080″
"Double Step"

Figure 5.5

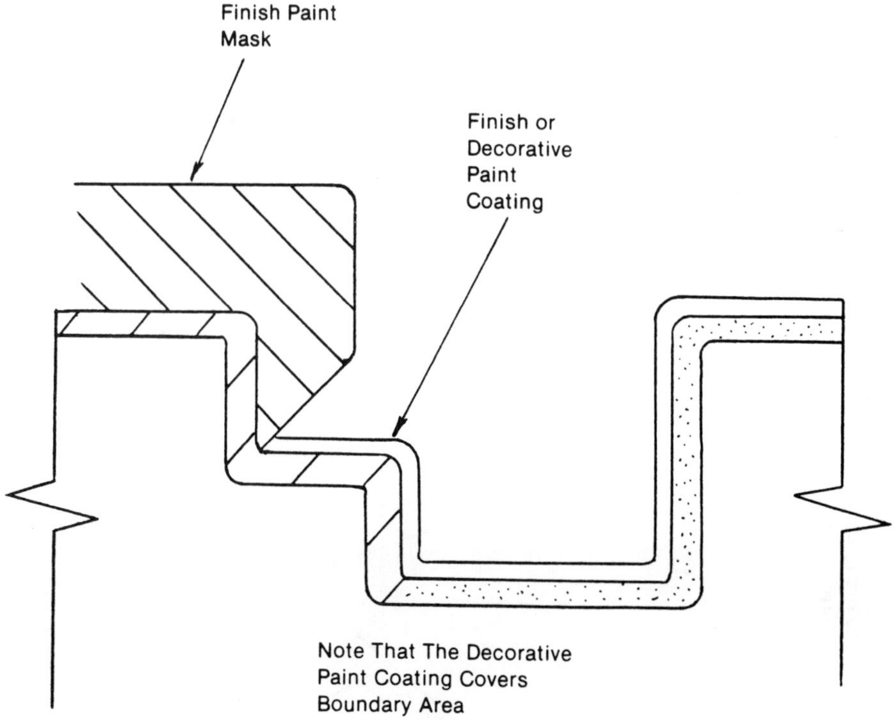

Figure 5.6

of vertical and horizontal bars. For very open grid work or on outer perimeters 12.7 mm (0.5 in) may be advisable. Recessing the step by this amount reduces the current density at the boundary sufficiently to avoid visual problems in this area. For the finish-paint coating the leading edges of the bars are capped back about 3.2 mm (.125 in) without a step, as shown in Figure 5.10, and the part is sprayed from the front with a decorative coat of contrasting finish paint. (See Figure 5.11.)

The primary reason for using selective plating in many automotive applications has been to insure good adhesion of decorative paint coatings on these parts.

A final consideration in the use of selective plating is the design of current-carrying paths. Since the whole part is not to be plated, special paths must be provided to carry the electrical current from the plating surface of the part to the contact tips on the plating rack. The final selection of the size and location of these paths is best left up to the manufacturer who also must design the plating rack for the part.

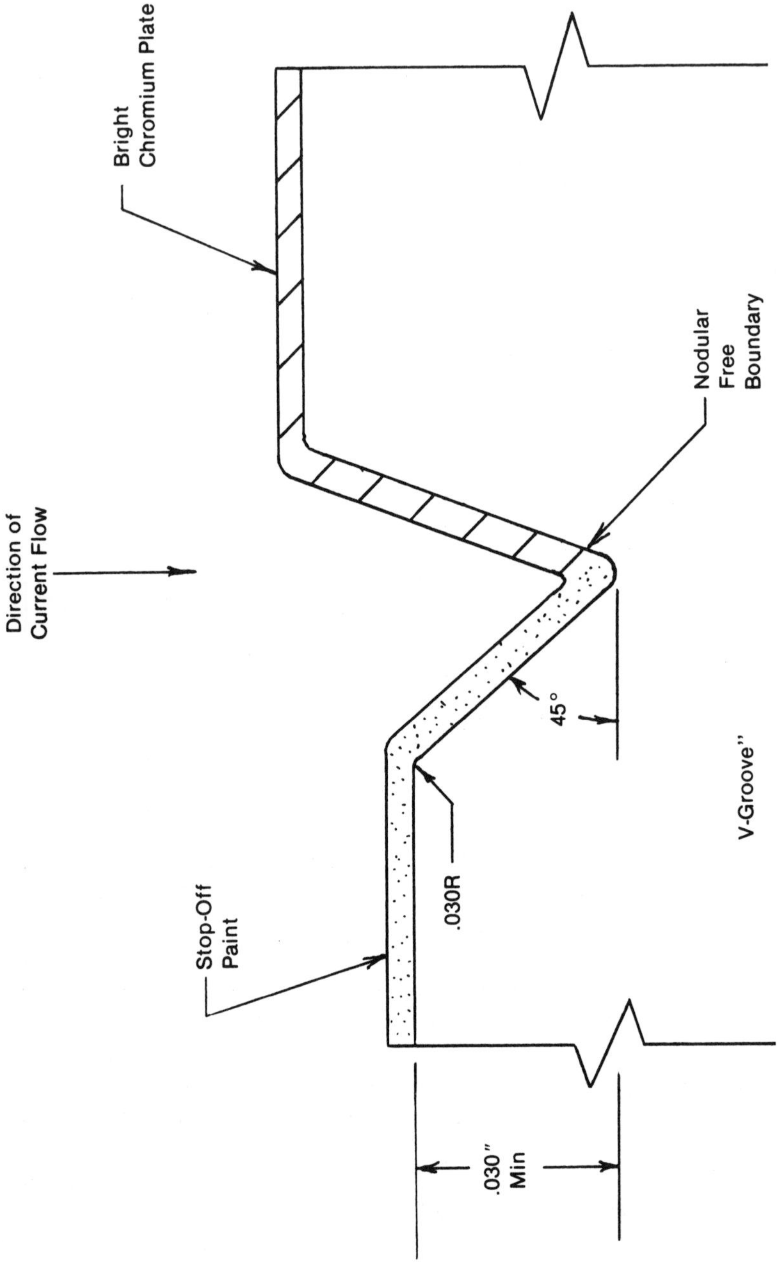
Bright
Chromium Plate
Direction of
Current Flow
Nodular
Free
Boundary
45°
V-Groove"
.030R
Stop-Off
Paint
.030"
Min

Figure 5.7

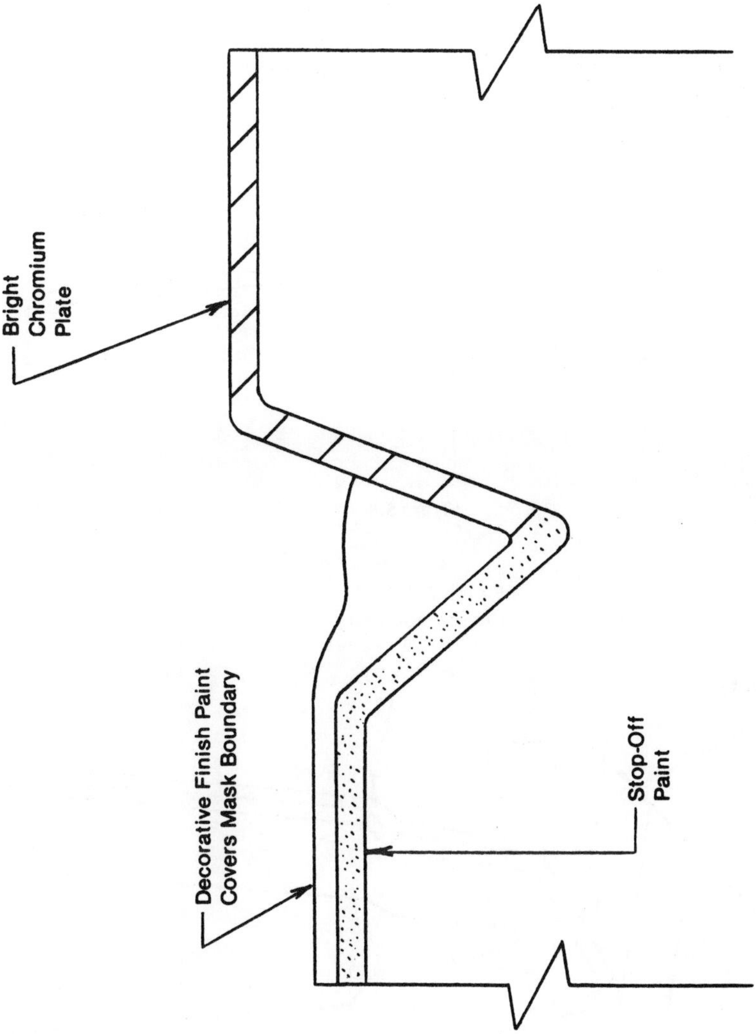
Bright
Chromium
Plate
Decorative Finish Paint
Covers Mask Boundary
Stop-Off
Paint

Figure 5.8

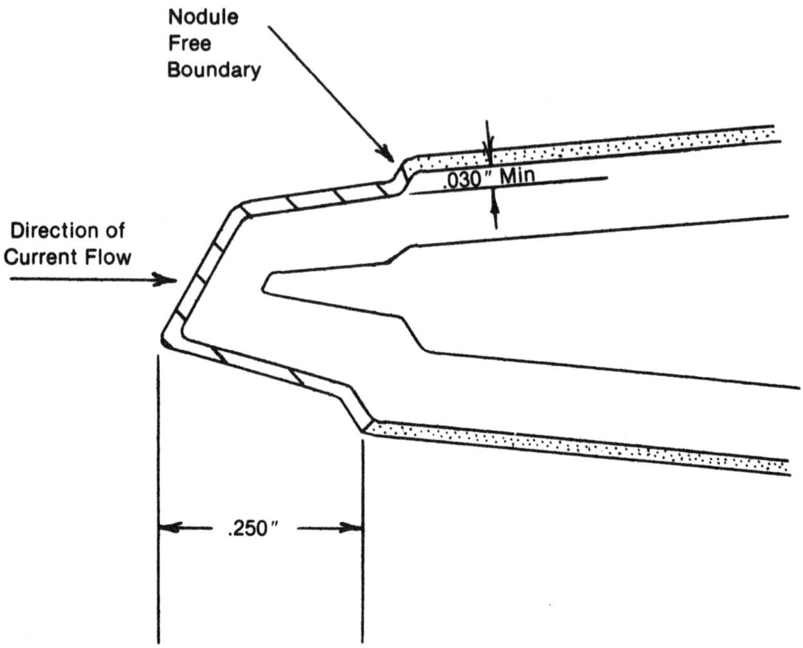

Figure 5.9

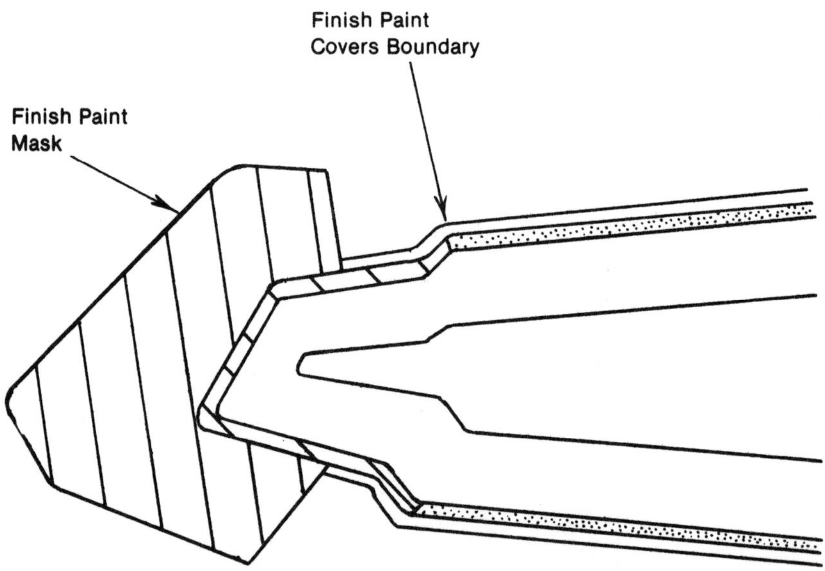

Figure 5.10

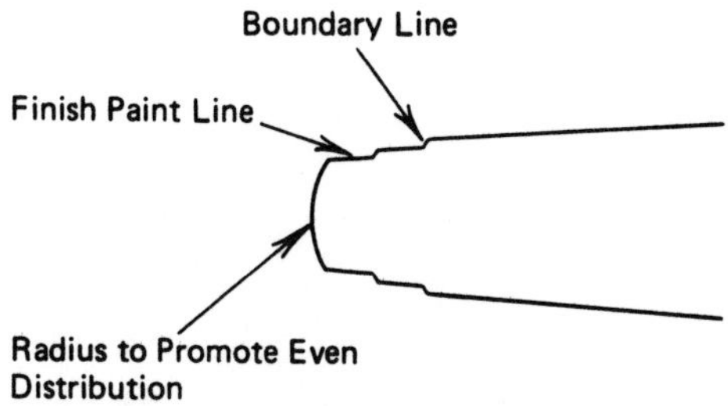
Boundary Line
Finish Paint Line
Radius to Promote Even Distribution

Figure 5.11

6

Rack Design Considerations

6.1 RACKS

Plated plastics offer the designer and stylist design freedom not heretofore obtainable. This design freedom, coupled with weight reduction realized with the use of plated plastic, has led to a steady increase in part size and complexity. With these larger and more complex plastic parts, the plastic plater is finding that the plating rack is one of the most important considerations. The proper design, construction, use, and maintenance of the plating rack contributes greatly to the quality of the part and to the plater's ability to meet production schedules.

Proper rack considerations can also greatly affect the plater's profitability. Guidelines for the design, construction, conditioning, use, and maintenance of plating racks are therefore provided.

6.2 ELECTROPLATING-RACK CONSTRUCTION

6.2.1 Plating-Rack Splines

Rack splines are the main body of the rack and carry the main current in the rack. Copper or aluminum are preferred for spline construction. Although copper has higher electrical conductivity than aluminum, see Table 6.1, aluminum weighs 30% of copper. To obtain the same electrical conductivity as copper, one would have to increase the cross sectional area of the aluminum by a factor of 1.75. This use of aluminum would reduce the weight of the rack by approximately 50% in comparison to the copper, and would carry the same amount of current as the

TABLE 6.1 Relative Conductivity Values

Metal	Amp/cm2	Amp/in2
Copper	155	1000
Aluminum	93	600
Brass	39	250
Nickel (9919)	39	250
Phosphor Bronze (Grade A)	28	180
Mild Steel	19	120
Lead	13	83
Titanium	4.8	31
Stainless Steel (302)	3.6	23
Stainless Steel (316)	3.4	22

copper rack. The reduction in weight gained by the use of aluminum would make the rack easier for personnel to handle and easier for machines to lift without sacrificing performance. Phosphorous bronze, brass and even steel are sometimes used when the superior conductivity of copper or aluminum is not required.

The generally accepted electrical conductivity of copper is 155 amp/cm2 (1000 amp/in2) of cross-sectional area. Steel is about 19 amp/cm2 (120 amp/in2), and stainless steel is only about 3.4 amp/cm2 (22 amp/in2). Table 6.1 shows the electrical conductivity of various metals.

The 155 amp/cm2 (1000 amp/in2) for copper may be exceeded up to 230 amp/cm2 (1500 amp/in2) for short periods of time, provided the member is immersed in the solution to provide dissipation of the heat generated. Improper material selection or insufficient cross-sectional area can cause less than the anticipated current delivered to parts, burn-off of the thin electroless deposit due to excessive heat buildup, and burning or blistering of the rack coating, which can cause solution entrapment and subsequent drag-out, solution contamination, and skip-plating.

6.2.2 Plating-Rack Hooks

The rack hooks which make the immediate contact with the cathode bar are best constructed as a continuation of the spline thus permitting maximum conductivity and substantial strength. Like the spline, the rack hook should be constructed of copper for conductivity reasons. The rack hooks must also be formed to conform as closely as possible to the design of the process cathode or work bar. This will allow for the maximum amount of current to be transferred through the rack and to the parts with a minimum loss of current due to poor or insecure contact areas.

The cross-sectional area of the hook or hooks must be sufficient to carry the total current required by the surface area of the parts to achieve the desired current density for the plating processes used: for example, the maximum current density for chromium is usually approximately 100–150 amp/ft2.

6.2.3 Rack Crossbars

The crossbar serves two purposes: to conduct current from the splines to the contacts and then to the parts and to give strength and rigidity to the rack. In the first case, mild steel is a satisfactory material when conductivity is not a consideration. A magnet is effective in determining through the coating, if steel crossbars or splines are present.

Generally bolts or rivets are used in joining rack cross members and splines. In assembly, riveted joints are tighter and more secure, yet bolts are often used for racks that will be involved in heavy production requirements, as bolts are much easier to remove when the rack is rebuilt.

Soldering all joints, whether bolted or riveted, is strongly recommended to prevent contamination from getting into the joint, to provide additional strength, and to enhance the joints ability to pass current. Soldering is essential when the crossbars are current carrying members.

6.2.4 Rack Contacts (Tips)

The function of the rack contact (or tip) is to hold the part in place and to provide electrical contact to the part. Vigorous air agitation in many of the plating baths calls for firm part placement on the rack, but too much tension on the contact may cause part distortion from the hot solutions in the plating process. Trial and error is usually used to determine the correct contact tension.

The use of corrosive solutions in the preplate system and the use of nitric-acid-bearing rack-stripping solutions has led to predominant use of stainless steel as a contact material. Titanium has also found some acceptance because of its excellent resistance to nitric acid rack-stripping solutions, but it cannot be soldered. This leads to two additional problems: (1) Stainless steel and titanium have approximately 2% of the electrical conductivity of copper, and (2) titanium has very poor elasticity compared to spring-tempered phosphor bronze. For these reasons, the length of the contact should be kept to a minimum, and if good spring action and heavier current-carrying capacity are required, a phosphor-bronze member should be used to lead to the contact with a short stainless steel or titanium piece used for the tip. Only the stainless steel or titanium should protrude from the coating.

6.2.5 Rack Coatings

Plastisol, the generally used rack coating, serves four purposes: (1) to protect the rack members from the corrosive plating solutions, (2) to inhibit plating of the rack by the preplate system, (3) to eliminate plating of the rack by the electroplating solutions, and (4) to insulate the contact tips to prevent robbing current from the part at the point of contact. Improper coating selection will lead to either deterioration of the rack or overplating of the rack coating, usually resulting in loss of the ability to chromium-plate the parts properly. It is extremely important that the coating be of the proper coating hardness or durometer. A too hard coating will become brittle and crack in a short period of time; a too soft coating will become porous and drag solutions from one tank to another.

The mechanism for controlling rack plating involves four steps in the preplate system. The best combination of time, temperature, and concentration must be determined for each of these steps in order properly to control rack plating:

1. Retention of chromium in the rack occurs in the etching process, Hexavalent chromium poisons the electroless reaction, thus preventing metal deposition.
2. Neutralization of the excess hexavalent chromium occurs without removal of the chromic acid absorbed in the coating.
3. Immersion in the catalytic solution (activator) must be controlled to assure proper absorption in the plastic without catalytic absorption in the rack coating.
4. Immersion in the accelerator must be controlled to remove excess catalytic material from the plastic and to remove any catalytic material from the outer skin of the rack coating.
5. The initiation rate of the electroless bath must be carefully controlled.

All plating racks must be sandblasted and primed before coating with plastisol. This improves coating adhesion and extends the life of the rack, particularly in the area of the exposed contacts. Loose coating at the rack contact tip permits hexavalent chromium to bleed out and run down into parts during the plating process, causing areas of skip-plating and scrap. For this reason, releasing agents to speed-up the trimming of the contact area should never be used. This is a major production problem to guard against.

All newly coated racks should be checked for proper cure and must be spark-tested for pinholes in the coating that are not visible to the naked eye. Any problem identified must be corrected before the rack is put into production.

6.3 ELECTROLYTIC-RACK DESIGN

A specific electrolytic rack is normally required for a specific part. Whenever a rack is used for a part for which it was not designed, resultant trade-offs in quality, performance, or productivity can be expected. Prototype racks should be rebuilt and tested with the initial part prototypes. The use of unusual equipment, such as auxiliary anodes and shielding, and the quantity of parts per rack must be anticipated at the time of job quotation.

Rack design can be optimized to improve the uniformity of deposit distribution and to avoid excessive buildup of deposits (burning or treeing) during electroplating. Uniform deposit thickness will minimize thermal cycle problems and replace-plating costs.

6.3.1 Part Placement

Thought must be given to part placement on the rack to optimize plate thickness distribution and visual quality and to minimize solution drag-out. Parts should be placed on the rack so that current density is balanced on the rack.

This can be accomplished by such means as spreading out parts in the center of the rack and setting center parts out while setting peripheral parts back. Place low-current-density parts toward tank anodes or add auxiliary anodes if needed.

Parts should be racked to keep shelf roughness on the bottom of the part in its service position. Parts should be placed to minimize drainage from one part onto another and to minimize solution carry-over from one plating tank to another. Parts should also be placed to keep gas from being trapped on significant surfaces.

6.3.2 Current Robbing

When part positioning alone will not suffice, another alternative is the use of current robbers (conductive members with a cathodic charge) placed in close proximity to the part and used to draw plate from the work piece. A simple example is shown in Figure 6.1. Robbers must be of stainless steel to resist

Figure 6.1

chemical attack during rack-stripping operations and should be positioned to reduce drag-out as much as possible.

6.3.3 Current Shielding

A third method of reducing plate buildup on high-current-density areas is to use current shielding. This is accomplished by placing a nonconductive member close to the part, thus selectively blocking the path the current must travel to get to the part.

In most cases, properly designed shields must be placed between the anode and the protruding high-current-density areas of the part and affixed to the rack securely. Clearance of 0.5 to 1.0 in is required between the shield and the part to allow solution access and rinsing. If proper design parameters are employed, shielding can be often as effective as robbing and has less maintenance costs due to the stripping and contamination problems associated with robbers.

6.3.4 Auxiliary Anodes

When the part design is such that it is impossible to obtain the desired thickness in a recessed area, auxiliary anodes are commonly used. There are three types of auxiliary anodes employed: soluble, insoluble, and bipolar.

Soluble auxiliary anodes have a positive charge and are placed in close proximity to the recess to be helped. Materials used for soluble auxiliary anodes for decorative plating are generally nickel, but in special cases other materials may be used.

These anodes are considered main auxiliary anodes when a single power source is applied to the anode or an extension of the anode that leads into the recessed area.

These anodes are considered secondary auxiliary anodes when a main rectifier supplies power to the regular anodes for plating the normal surfaces and a second rectifier supplies power to the auxiliary anodes for plating the recessed areas. In this case, the auxiliary anode is usually mounted on the plating rack but insulated from the cathodic portion of the rack. When this approach is used, soluble anodes must be added after the preplate and removed prior to chemical stripping of the rack contacts.

Anodes that pass current but do not dissolve are called insoluble anodes. Insoluble anodes must be used somewhat differently from soluble anodes. Insoluble (nonsacrificial) anodes can require (1) a slightly higher electro potential at the anode to pass the same current, (2) metal replenishment with metallic carbonates to control the drop in pH that the use of these anodes causes and to replenish the metal plated from solution by these anodes, (3) more frequent solution treatment, since certain types of electroplating brighteners may be adversely affected by the use of insoluble anodes, resulting in higher brightener consumption, and (4) closer pH control, due to depletion of metal ions.

Insoluble auxiliaries are generally secondary anodes (attached to the plating rack and powered by separate rectification). Due to the corrosive nature of some preplate solutions, the insoluble auxiliaries are sometimes mounted after the preplate system. Most insoluble auxiliaries, however, are left in place through the entire preplate and electroplate cycle.

Several materials have been used for insoluble auxiliary anodes, including iron (chromium or gold plating only), lead (chromium or gold plating only), stainless steel (chromium or gold plating only), and platinized (platinum-coated) titanium. Platinized titanium is most widely used, even though it is the higher in cost, because it offers long life and minimizes problems with the electroplating solutions.

Most recently, titanium coated in palladium/nickel is being used in place of platinized titanium. The titanium/palladium nickel material offers much greater life than the platinized titanium and reduces even further the problems commonly associated with the electroplating solutions, namely solution contamination.

The principal of the bipolar auxiliary anode is to use an anode without an external current hookup. The current is directed toward a low-current-density area or away from a high-current-density area via the bipolar effect. The bipolar effect is simply that a metallic object, when placed in a current-carrying solution, will set up a positive charge in the vicinity of the cathode and a negative charge in the vicinity of the anode.

6.3.5 Part Contact

The number of contacts required for plastic parts is greater than the number required for metal parts for several reasons: (1) Stainless steel has a very limited current-carrying capacity; (2) plastic is easily distorted, and limited pressure may be applied to any one point; (3) thin electroless deposits have a limited current-carrying capacity; and (4) intricate designs may require unusual racking to get proper bright-plate throw.

Table 6.2 relates surface area to the suggested number of contacts.

The following general recommendations apply to contact design for plastic:

Contact the part in a medium to low-current-density area.

The contacts should be sloped back to the frame to prevent dripping on parts below.

The contact tip should be made of stainless steel, titanium, or phosphorous bronze-stainless steel or titanium combination.

The contacts should not lose tension after repeated loading.

TABLE 6.2 Required Contacts

Square Decimeters Plated	Square Feet Plated	Number of Contacts Required
0–2.32	0–¼	1
1.55–4.65	¼–½	2
4.65–9.29	½–1	4
6.97–18.28	¾–2	6
13.94–27.87	1½–3	8
23.23–37.16	2½–4	12
27.87–55.74	3–6	16

The contacts should not shade a significant area of the part.

The contacts should not come out of adjustment after repeated loading.

The contacts should be relatively easy for the operator to load.

Use enough contacts to carry the current adequately, even if one is lost as a current source.

Spring contacts should be designed with a sufficient radius on bends to prevent weakening and fracturing of the spring.

Spring contacts should be designed so that the bend closes when flexing and should allow for adjustment of contacts while in service.

Orient sharp points and corners of the part to the center of the rack.

Areas of the part where contact is made must be properly supported to avoid part distortion.

Avoid placing a contact in a blind hole or deep recess as burn-off or skip-plating may result.

Rounded point contacts should be used instead of flat contacts to avoid skip-plating from poor rinsing.

Contacts should not be placed on high-quality surfaces because of the possibility of misplating or discoloration in the contact area.

All contacts must be free of plastisol and primer in the area where they contact the part. If the contact areas are not clean, bipolar skipping can occur.

6.4 PREPLATE RACKING

The function of the plating rack for preplating is different from the required for electroplating. All that is required for preplate is that the parts be secured; the racks need not carry electric current to the parts. For this reason, special bulk or basket type racks can be used that are designed to hold as many parts as possible, with special consideration given to solution drainage. The electrolytic racks can also be used.

Using the bulk method, small parts can be processed in polypropylene barrels. Larger parts or parts that do not lend themselves to barrel plating can be processed in layered baskets of 316 stainless steel or titanium. Some current solutions will not tolerate outside metal content and therefore the stainless baskets may require coating. Baskets should be built to contain no pockets or blind holes. All basket welds must be deburred and ground smooth. Titanium and stainless steel baskets can be spot welded or heliarc welded. The stainless steel baskets may be coated to prevent chromic, hydrochloric and nitric acids from attaching the various components of the stainless steel and then depositing the metals into the bath. If not coated, the stainless steel will become rough and this roughness will damage the parts during processing. Coating offers the advantage of providing a cushion for unsecured parts, yet the preplate solutions will plate the entire coated surface.

No consideration need be given to current distribution. The parts are racked as closely as possible, the only concern being that of drainage. Preplate systems work effectively when heavily loaded. Four to eight times as many parts may be processed in a given tank volume by the use of bulk preplating. Since current is not used, the parts are generally contacted in recessed areas, holes, or other hidden areas. This offers a virgin contact for the electroplating.

When using "through-racking," the term used to describe racking that is common to both preplate and electroplate, the positioning of parts is dictated by the requirements for electroplate distribution. Frequently a compromise is made to obtain acceptable common part positioning for preplate and electroplate. The number of parts per rack, or plating tank volume, is greatly reduced. Since parts are processed straight through, the contact must be suitable for both preplate and electroplate. It must contact in a medium- to high-current-density area and afford good rinsing. Figures 6.2 and 6.3 illustrate the two methods of racking for preplate.

6.5 RACK CONDITIONING

Prior to using a new stainless steel or titanium bulk preplate rack in production, coated or uncoated the welded joints must be inspected to assure that they are relatively pore-free.

If the electroplate rack is to be used for preplating, the rack should be run through the rack-stripping solution to remove some of the surface plasticizers. This helps insure proper chromium retention in the rack coating for control of rack plating. The same effect can be achieved by soaking new racks in chromic acid etch solution prior to their use in production.

Figure 6.2 Semi-bulk preplating method

6.6 RACK STRIPPING

In bulk preplate systems in which electroless copper is the preplate metal, the rack will be automatically stripped during the preplate cycle if a chromic acid and sulfuric acid etch is used. Otherwise, as with electroless nickel, a nitric acid dip may be used.

The electroplate buildup on the contracts of electroplating racks can be considerable and should be removed after each cycle to insure clean positive contact to every part. Most installations strip their racks in various nitric acid based strippers after each use. This practice does, however, have a deleterious effect on the rack coating and helps shorten rack life. Some installations use a reverse current strip that contains sulfuric acid and other proprietary

Figure 6.3 Straight through cell plating method

chemicals. If this strip process is used, titanium contacts must be used in place of the stainless, as the reverse current process will violently attack all forms of stainless steel.

6.7 RACK CARE AND STORAGE

The rack coating should be regularly checked for cracks or holes and promptly repaired before acids deteriorate the interior of the rack. Correct adjustment of the contacts should also be inspected to insure the parts will process without distortion.

Proper storage should be provided for hanging racks when they are not in use to prevent damage from stacking them together.

7

Electroless Plating Considerations

7.1 INTRODUCTION

Electroless plating (preplating) of plastics is a chemical process, consisting of several steps that promote adhesion, catalyze, and deposit a thin, adherent layer of metal onto a plastic substrate. This metal coating can subsequently be electroplated much like a completely metal part.

The proper selection and control of this process is critical to maintain a product with good adhesion, good corrosion resistance, and acceptable appearance. The primary factors influencing the selection of the electroless plating process are the specific resin to be plated and the end use requirements of the plated product.

7.2 THE MECHANISM OF ADHESION

Adhesion is normally generated between the plastic substrate and the electroless metal by use of an etchant, or etchant system. The etchant removes components of the plastic such as low molecular weight portions, filler materials, or selective resins, leaving a "micro-porous" surface on the plastic part. The roughened surface created by the etchant provides sites for mechanical bonding, a physical interlocking of plastic and metal. The greatly increased surface area also allows the creation of many more sites for chemical bonding, a chemical interaction between plastic and metal. The resulting combination of the chemical and mechanical bonds formed in the etching process provides a finished part with good tolerance to the chemical and natural forces (adsorption of moisture and/or gases, corrosion, thermal expansion and contraction, etc.).

7.3 COMPOSITION OF PLATEABLE PLASTICS

Most plateable plastics are a mixture of compounds. These mixtures are formulated to allow the etchant system to selectively or preferentially remove one portion of the compound. ABS, for instance, is a mixture of two copolymers - acrylonitrile styrene and butadiene stryene - while modified polyphenylene oxide is a mixture of PPO styrene and butadiene styrene. The plateable polypropylenes are a mixture of high-molecular weight and low-molecular-weight polypropylene.

Other types of plateable resins contain fillers which can be removed by an etchant. Plating grades of polysulfone, nylon, and some grades of polypropylene are examples. The filler in these plastics is dissolved or otherwise removed from the surface of the plastic part leaving the "micro-porosity" necessary for the development of adhesion.

Because plateable grades of plastics are a mixture of compounds, including lubricants, antioxidants, flowing agents, mold release compounds, it is important to realize that there is no such thing as a uniform surface throughout an injection molded part. It is due to the very dynamics of the injection molding process that the surface of the molded part is formed under different conditions from one area of the part to another. These varying surfaces in the molded part require different conditions in etching for optimum adhesion and final appearance. For this reason, multi-step etchant systems, wherein the different steps in the system are tailored to react with the varying conditions on the surface of the molded part, often yield better overall performance than single-step etchants. The use of a single step etchant system often requires sacrificing some adhesion in the harder to etch areas and sacrificing some of the appearance in the easier to etch areas. For this reason, the plater who wishes to optimize adhesion will be well advised to investigate all types of etchant systems, including those with pre-etch conditioner steps.

7.4 ELECTROLESS PREPLATING PROCESSES

7.4.1 Electroless Plating Chemistry

The chemistry of the electroless preplating process can be divided into three general areas: Adhesion promotion, catalyzation, and electroless deposition. The adhesion promotion process normally consists of an etching step or steps, and may also utilize pre-etch conditioners and neutralizers. This portion of the process not only affects the ultimate adhesion and appearance of the part, but also the catalyzation and electroless deposition portions of the process. Optimization of this process area ensures the adhesion behavior of the final end product. If oxidizing type etchants (chromic acid) are used, reduction of the oxidant is required before further processing. The products used for this purpose are usually referred to as neutralizers. Additional additives or process steps can be utilized in this portion of the process to improve the catalyzation of certain plastics.

Catalyzation is achieved by use of a catalyst and an accelerator. The catalyst provides a catalytic layer, generally palladium based, on the etched plastic surface. The accelerator improves the uniformity of the catalyzed surface by removing a portion of the tin normally associated with the palladium catalysts, and also strips any unwanted catalyst from rack coatings or plating resists.

The electroless deposition step is the actual deposition of the thin metal film which serves as the basis for electroplating. This step is normally either copper metal (electroless copper) or a nickel and phosphorus alloy (electroless nickel).

7.4.2 Electroless Plating Processes

Electroless preplating is commonly performed in two basic modes, straight-through and semi-bulk. The straight-through process ties the electroless preplate and the electroplating processes together into one continuous system. This technique utilizes one rack for both processes and may be used with either electroless nickel or electroless copper for plating ABS and some other resins. The semi-bulk processing technique uses separate fixtures for electroless preplating and electroplating. Electroless copper is normally used, and the resins plated are not limited by rack plating problems.

Both techniques have advantages and limitations. Typically a straight-through process can produce a part at less cost due to the savings in handling and handling scrap, and has a built in capability to handle parts with plating resist (stop off paint). The process is limited because it requires a large volume of parts of the same resin (or very similar resins), has critical process control parameters, and has very limited rework capability.

The semi-bulk technique has the advantages of a more diverse selection of resins, the ability to offer multiple finishes on the same parts, and offers some rework capability. The process also allows parts to be racked in the electroless preplating process without the normal electroplating constraints, thereby allowing more dense part distribution on the rack, orientation for best drainage, etc. The disadvantages to semi-bulk processing are the extra labor, tooling, and handling.

7.5 SELECTION OF A PROCESS

Due to the complex nature of the preplate system, which involves many interactions of the various chemical steps, it is recommended that a complete system be chosen as recommended by the chemical supplier, with any substitution approved by the supplier.

In making a selection of a process, it is recommended that tests be run with representative moldings of your end product using the resin selected that will best meet the requirements. The Peel Test described in Chapter 10 may be applied. Using the complete system selected, process the parts, adhering closely to recommendations of time, temperature, and concentration. In the event that one is unable to run these tests in-house, the supplier may perform them. Evaluation tests performed in-house will result in a better understanding of how the process performs. These moldings should be selected from a single controlled molding run of plating-grade plastic. These preplated parts then must be electroplated as follows:

First, assuming that chromium is to be the final finish and recalling that the preplate system is what one is interested in evaluating, no less than twelve parts should be electroplated with a "standard" composite. A suggested composite is:

17.8 m (.0007 in) Bright Acid Copper

7.6 m (.0003 in) Bright Nickel

0.25 m (.000010 in) Chromium

Second, at least four additional parts from each cavity should be electroplated with 38 m (.0015 in) minimum bright acid copper only for use in adhesion-pull testing. Experience has shown that with high-adhesion levels, the plate often tears and the specimen is lost; therefore, greater thicknesses of copper may be necessary. Increased thicknesses of copper will influence the adhesion values upward, so all tests should be conducted with the same thickness. Studies have shown that ABS parts require four weeks' aging at room temperature before they attain their maximum adhesion, and other plastics require more time. Similar aging results may be achieved by heating the parts in an oven for four hours at a given temperature within the range of 71–82C (160–180F). To insure uniformity of aging, pull-test specimens should be subjected to accelerated aging. They should then be tested for adhesion by the standard peel test outlined in Chapter 10.

Suggested minimum adhesion values for acceptance of a preplate process should be as follows:

ABS	Minimum 5 lbs. (Range 5–20 lb)
Modified PPO	Minimum 4 lbs. (Range 4–15 lb)
Filled Polysulfone	Minimum 2 lbs. (Range 2–8 lb)
Filled Nylon	Minimum 2 lbs. (Range 2–8 lb)
Filled Polyester	Minimum 2 lbs. (Range 2–8 lb)
Polycarbonate	Minimum 2 lbs. (Range 2–15 lb)
Polyetherimide	Minimum 4 lbs. (Range 4–30 lb)

Required minimum adhesion values will vary with the plastic substrate and with the application. As the physical properties of the resin become more "metal-like," the required metal-to-plastic bond becomes less. This is because the loss of adhesion caused by bond failure due to the difference of expansion (and contraction) rate between the plastic and the plated metal becomes less likely as the coefficient of expansion of the plastic becomes closer to that of the plated metal. Also, as the heat-deflection temperature for the plastic becomes a higher value, the plastic is less likely to "move" during and after plating; thus, less force is applied to the metal-to-plastic bond. Of course, there is a minimum bond required, depending on what forces the end product will be expected to encounter in use. One reference point is perhaps that the typical adhesion of plating to zinc die casting is in the range of 2.27–15.88 Kg/2.54 cm (5–35 lb/in). The lower level is common on zinc die castings that show no adhesion failure in the field.

7.6 EVALUATION OF THE SYSTEM

The copper-nickel-chromium-plated parts should be tested by thermal cycling for at least three cycles (see the Thermal Cycle Tests in Chapter 10). After the cycle test the parts should be checked carefully for blisters, loss of adhesion, and cracking. Blisters and loss of adhesion from the plastic may be an indication of the inability of the preplate system to provide uniform adhesion over the entire part. Cracking usually indicates highly stressed moldings.

Properly selected preplate systems will provide adhesion even in problem areas, such as gate areas and hard-to-etch edges. Poorly etched parts preplated in electroless nickel preplate systems are subject to loss of adhesion when moisture and corrosive atmospheres penetrate to the substrate.[1] This penetration can take place very quickly in areas of misplating, such as in contact areas. Minor thermal cycle failure, such as fine hairline cracks, allows rapid moisture penetration to the bond area between the metal and the plastic. For this reason, it is recommended that after parts are thermal-cycled, they be subjected to accelerated corrosion testing, if normal duty for the parts will cause it to be exposed to corrosive environments, as with exterior automotive or marine applications.

Normally, exposure of 48 hours to the CASS Test (ASTM B-368) is a good test of corrosion resistance. Any loss of adhesion on corrosion penetration should be judged as a failure of the preplate system.

In addition, the following items should be considered in selecting a process:

1. Cost per square meter or square foot of chemicals (start-up and maintenance).

[1]R. G. Wedel, "Characteristics of Corrosion-Associated Blisters of Plated Plastics," *Plating*, January, 1975, and "A Mechanism for Corrosion-Associated Blisters on Plated Plastics," *Plating*, March, 1975.

2. Appearance of finished product.
3. Adaptability of various plastic materials to plating.
4. Rack plating.
5. Solution stability to contamination.
6. Ability to process stop-off.
7. Latitude and ease of operation.
8. Ease of analysis.
9. Electroplate composition and thickness.
10. Adaptability to existing equipment.
11. Waste treatment techniques and costs.
12. Overall costs.

7.7 PROCESS CONTROLS

After a system is chosen, carefully review with the vendor all the steps in the cycle in order to understand the purpose of each step. Solution controls, temperature controls, time controls, rinsing, rack maintenance, and part or solution contamination are important factors in the overall reliability of the system and must be given extensive consideration.

7.7.1 Solution Control

Set up solution controls as recommended by the vendor. Many solutions will require daily chemical analysis, depending upon the size of tanks and production volume. Records should be kept that show the analyses, the amount of surface area that has been processed, and the additions which have been made. In addition, statistical methods may be used to monitor, control, and reduce variation in the process tanks.

7.7.2 Temperature Control

The rate of all chemical reactions change with temperature, some more than others. The vendors' temperature recommendations should be followed. Automatic temperature control should be used where necessary. Check and record temperatures on a regular basis, including those of "room temperature" baths.

7.7.3 Time Control

If at all possible, the preplate cycle should be automated. A machine will provide uniform immersion and rack-transfer times, producing a product more consistent in quality.

The vigilant use of timers will considerably aid the control of nonautomatic systems. Records should be kept of all cycle changes.

7.7.4 Rinsing

Thorough rinsing is important in all plating operations, but in plastic preplating, rinsing is a very critical step in the process. The etched surface of the plastic acts somewhat like a sponge, requiring multiple and thorough rinsing to clean the surface of parts and to prevent contamination of subsequent process steps.

7.7.5 Rack Design and Maintenance When Electroplating Racks Are Used for Preplating

Racking plastic for electroplate requires racks with more contact points and a higher degree of rack maintenance than when plating on metal parts. Rack contacts should generally be chemically stripped of plating buildup after each pass through the electroplating cycle. Special attention should be paid to rack contact tension due to the ease with which plastic parts can be distorted. Solution carry-over in split and cracked rack coatings may cause skip-plating and shortened solution life. (See also Chapter 6.)

7.7.6 Rack Design and Maintenance for Semibulk Preplating

Trays should be designed to hold the parts for best drainage; often trays may be designed to hold the parts loose so as not to cause warpage. Trays should be made from 302 or 304 stainless steel and kept in good repair. PVC coating and titanium are sometimes used for part of the tray to reduce abrasion between the tray and a critical plastic surface.

7.7.7 Contamination

Contamination of parts and solutions should be minimized by controlling immersion time, rinsing, drag-over, and filtration. Controls must provide for sufficient immersion time for proper neutralization and rinsing. Rinse quality must be maintained by providing good-quality water at a sufficient flow rate. In some cases, drag-over should be controlled by providing drain-off time over the process station and by avoiding solution entrapment caused by improperly racked parts or poor rack maintenance. Finally, proper filtration is required to remove unwanted solids from preplate solutions.

Part contamination must be avoided by handling parts with gloves during processing. Contamination of parts from airborne oil and chemical mists must be avoided by providing adequate ventilation of plating solutions. Chromic acid-bearing mists from chrome plating operations are particularly harmful and must be scrupulously avoided.

8

Electroplating Considerations

8.1 BASIC CONSIDERATIONS

The selection of plate composition, thickness, and the process for electroplating is the first consideration in producing an electroplated part. The cost of the electroplating process varies widely, as the demands can vary according to the use and environment in which the end product must function. If a product is electroplated for purely decorative purposes and its use is in a dry or moderate climate, plate composition and thickness can be minimal. If, on the other hand, the product is used in an environment that would subject it to both physical and corrosive abuse, the plate composition and thicknesses must be more extensive and costly. Minimum protection is required, for example, on a nameplate on a household appliance, while maximum protection is required on an automotive grille or wheel cover.

After electroless preplating, plastic may be electroplated with a wide variety of metal coatings for many reasons other than decorative. Plastic may also be plated for electrical conductivity, physical-property improvements, and electromagnetic shielding. Because decorative chromium plating is by far the largest application in electroplated plastics, it will be the subject of this discussion.

8.2 MINIMUM ELECTROPLATING REQUIREMENTS TO PRODUCE A BRIGHT CHROMIUM FINISH

Due to the fact that plastics possess a higher thermal coefficient of expansion than electroplated nickel-chromium coatings, it has become common practice to use a bright acid electroplated copper plate as a base for the nickel-chromium plating.

The copper deposit has two important functions: (1) copper plated from acid solutions is capable of a high degree of leveling, which performs the important function of leveling and brightening the etched electrolessly coated plastic base, and (2) the copper deposit is relatively soft and ductile and has the ability to expand and contract, acting as a buffer between the plastic substrate and the overlying nickel-chromium deposit.

Chromium has the advantage of being a hard, highly tarnish-resistant metal that will retain its luster when deposited over a bright substrate in low thicknesses. Unfortunately, chromium is very highly stressed and lacks the ability to expand and contract without fracture and, therefore, produces an unsatisfactory finish when plated directly over easily corroded substrates such as copper. Nickel, although not as noble and tarnish-resistant as chromium, is much more noble and tarnish-resistant than copper. It is the nickel deposit, along with the chromium, that provides the corrosion protection of the underlying copper.

8.3 SERVICE REQUIREMENTS

8.3.1 Very Mild Service Requirements (SC 0)

Very mild service normally requires an electroplate composition with the following minimum thicknesses:

Bright Copper	10 μm (0.4×10^{-3} in)
Bright Nickel	5 μm (0.2×10^{-3} in)
Chromium	.13 μm (0.005×10^{-3} in)

8.3.2 Mild Service Requirements (SC 1)

Mild service normally requires an electroplate composition with the following minimum thicknesses:

Bright Copper	15 μm (0.6×10^{-3} in)
Bright Nickel	7 μm (0.3×10^{-3} in)
Chromium	0.25 μm (0.01×10^{-3} in)

Because there is a relationship between the amount of copper (the most ductile deposit) and the overlying nickel-chromium, for thermal cycle performance, it is appropriate to maintain a ratio of copper to nickel of at least 1 to 1. When increasing the nickel and chromium thicknesses, it may be necessary to increase the ratio of copper to nickel to 2 to 1.

This service condition normally includes exposure to warm, dry atmospheres, with the coating subject to minimum abrasion.

8.3.3 Moderate Service Requirements (SC 2)

As the service requirement increases, the amount of bright nickel should be increased. The minimum thicknesses recommended for moderate service requirements is as follows:

Bright Copper	15 μm (0.6×10^{-3} in)
Bright Nickel	15 μm (0.6×10^{-3} in)
Chromium	0.25 μm (0.01×10^{-3} in)

The minimum recommended thickness of bright nickel can be reduced to 10 μm (0.4×10^{-3} in) if microdiscontinuous chromium (see following discussion) is used.

Further improvements in corrosion protection (protection of the underlying copper) may be achieved by incorporating two additional systems:

1. *Duplex Nickel.* Duplex nickel deposits retard corrosion penetration by utilizing a sulfur-free, semibright plate under the bright nickel plate. When corrosion occurs (at any discontinuity in the chromium top coat), penetration through the bright nickel layer to the semibright layer produces a corrosion cell that allows the more active bright-nickel layer to corrode laterally rather than allowing penetration through the more noble semibright layer. For the same thickness, duplex nickel has an added advantage over all-bright nickel only in that the semibright nickel layer is more ductile. This adds to the cushion effect of the copper.

2. *Microdiscontinuous Chromium.* Microdiscontinuous chromium retards corrosion penetration through the underlying nickel deposit by exposing a greater percentage of underlying nickel through micropores or microcracks. As the pore count or crack frequency increases in the chromium deposit, the penetration rate through the underlying nickel decreases. The two most common methods for providing microdiscontinuous chromium are:

   a. A particle-containing predip or nickel process wherein inert particles are incorporated into a thin bright-nickel deposit. These particles promote pores in subsequent chromium deposits.

   b. An induced-stress system in which a thin layer of nickel-containing stress-producing agents is utilized between the bright nickel and the chromium. These stressed-in nickel deposits produce a microcrack pattern in the chromium layer.

NOTE: Chromium deposits plated from trivalent chromium baths are microdiscontinuous.

This service condition includes indoor exposure where condensation may occur—kitchens and bathrooms, for example. Certain automotive specifications call for this coating for some components.

8.3.4 Severe Service Requirements (SC 3)

For severe service requirements, it is recommended that a duplex nickel system be combined with the microdiscontinuous chromium system with the following minimum thicknesses:

Copper	15 μm (0.6×10^{-3} in)
Semibright Nickel	15 μm (0.6×10^{-3} in)
Bright Nickel	5 μm (0.2×10^{-3} in)
Microdiscontinuous Chromium	0.25 μm (0.01×10^{-3} in)

It is also acceptable to use a duplex nickel system with regular chromium deposit. The thicknesses shown below apply in that case:

Copper	15 μm (0.6×10^{-3} in)
Semibright Nickel	19 μm (0.75×10^{-3} in)
Bright Nickel	6 μm (0.25×10^{-3} in)
Chromium	0.25 μm (0.01×10^{-3} in)

This service condition includes exterior exposure where occasional or frequent wetting by rain, dew, or strong cleaning solutions can occur—for example, lawn furniture, bicycles, and hospital furniture.

8.3.5 Very Severe Service Requirements (SC 4)

For very severe service requirements, it is recommended that a duplex nickel system be combined with the microdiscontinuous chromium system with the following minimum thicknesses:

Copper	15 μm (0.6×10^{-3} in)
Semibright Nickel	19 μm (0.75×10^{-3} in)
Bright Nickel	6 μm (0.25×10^{-3} in)
Microdiscontinuous Chromium	0.25 μm (0.01×10^{-3} in)

With respect to corrosion performance, it is also acceptable to use a duplex nickel system with a regular chromium deposit. The thicknesses shown below apply in that case:

Copper	15 μm (0.6×10^{-3} in)
Semibright Nickel	23 μm (0.9×10^{-3} in)
Bright Nickel	7 μm (0.3×10^{-3} in)
Chromium	0.25 μm (0.01×10^{-3} in)

This service condition includes exterior exposure under conditions encountered by exterior components on automobiles and by marine hardware in salt-water service. Certain automotive specifications call for an SC 2 coating for some components.

8.4 CONTROL OF SOLUTIONS

The ability to pass thermal cycle requirements depends in part on maintaining the ductility of the electrodeposited layers. The ductility of the deposits may be monitored by performing the Micrometer Bend Test for Quality (ASTM B-490), using the exterior of the foil, as plated, as the outside of the bent foil.

8.4.1 Interpretation of Tests on Nickel Foils

A properly maintained semibright nickel will have a value of 100%. A properly maintained bright nickel will have a value of 15% or better. It should be recognized that since the bright nickel is plated over relatively smooth copper surfaces when plating on plastics, the nickel can be operated with less brightener than is normally used when plating on other substrates, thus promoting ductility of the bright nickel.

8.4.2 Variation of Tests for Copper Foils

Since copper is more ductile than nickel and some apparently ductile copper deposits yield poor thermal cycle results, the Micrometer Bend Test may be modified when judging the ductility of the bright acid copper deposits. The foil is prepared as for nickel. The thickness of the foil is determined. The foil is bent over and crushed between the jaws of a micrometer until the reading on the micrometer is just twice the thickness of the foil. If the foil can then be straightened out flat without breaking, it is judged to have a ductility value of 100%. If the foil breaks on straightening, another strip must then be tested. The second strip is crushed a little less between the jaws of the micrometer (higher reading on the micrometer scale) and again straightened out. This procedure is continued until a micrometer reading is determined at which the foil does not break on straightening.

This reading is then divided into the thickness to give the factor. Properly maintained copper will have a value of 100%.

Factors that cause poor ductility in plating baths are (1) excessive addition agents, as well as their improper balance, (2) organic contamination, (3) metallic contamination, and (4) controlled anions (chlorides in copper). When ductility factors drop below those given, the solutions should be appropriately treated to correct the condition.

8.4.3 Solution Analysis

All plating solutions should be analyzed on a regular basis. Records should be kept that show the analyses, the number of ampere-hours (or equivalent, e.g., number of racks) used per shift, and the additions made. Ampere-hour brightener feeders are recommended. In addition, statistical methods may be used to monitor, control, and reduce variation in the process tanks.

8.4.4 Temperature

The temperature of all cleaning and plating baths should be thermostatically controlled. Record temperatures regularly.

8.4.5 Time Control

If at all possible, the electroplating cycle should be automated. A machine will provide uniform process control. Immersion times and rack-transfer times must be controlled.

The vigilant use of timers will aid considerably the process control of nonautomatic systems. Record any cycle changes.

8.4.6 Rinsing

Thorough rinsing is important in all plating operations. Provide sufficient rinses and control the water quality. Use deionized water prior to each electroplating bath and for final rinsing.

8.4.7 Current Controls

Provide current controls for voltage- or amperage-controlled power sources. Design racks and work carriers to operate at a uniform current density. The proper control of current density will greatly assist in obtaining the desired electroplate thicknesses and appearance. Ampere-hour meters can be used to control brighteners.

8.4.8 Rack Maintenance

Rack design and maintenance are important when plating on plastic. Generally, more contact points are required than when plating on metal. Contacts should be placed as close to the high-current-density areas as possible, and the tension must be controlled to prevent distortion of the plastic part. Exposed, uncoated rack areas are to be avoided to eliminate treeing in the plating baths. These "trees" may flake off, causing shelf roughness. Provide for the stripping of racks after each round.

8.4.9 Contamination

Solution contamination can be controlled by the following means:

Time Control: Provide sufficient immersion time for treatment, neutralizing, and rinsing steps.

Rinsing: Provide sufficient rinsing volumes and control water quality. Spray techniques and air agitation may be beneficial.

Drag-over: Control transfer rate. Avoid solution entrapment that may be caused by improperly racked parts or poor rack maintenance.

Parts can be contaminated before and during processing. Control the following:

Handle parts carefully during storage and racking. Avoid fingerprints (i.e., use gloves) and shop dirt.

Avoid airborne oil and chemical mists. Vent tanks where applicable (e.g., chromium-plating tank).

8.5 TESTS CONDUCTED ON FINISHED PARTS

8.5.1 Thermal Cycle Test

The thermal cycle test is a test that will check both the preplate and the electroplate. Nonfailure will indicate that the adhesion, ductility, and plate construction are adequate for the part tested. It is a useful practice to keep a supply of the reference parts on hand to control for system imbalance and to run several on each shift for thermal cycle testing.

The thermal cycle tests are described in Chapter 10.

8.5.2 Accelerated Corrosion Tests

There are several accelerated corrosion tests that may be used to check the durability of the electrodeposited coating; for example, the Salt Spray Test (ASTM B-117), the Acetic Acid Salt Spray Test (ASTM B-287), the Corrodkote Test (ASTM B-380), and the CASS Test (ASTM B-368). Of these tests, the CASS Test is preferred, since it appears to correlate better to actual service requirements. It is not uncommnon for a part to pass a thermal cycle test only to fail badly with loss of adhesion after a CASS Test. The thermal cycle and CASS corrosion tests may be combined (see Chapter 10).

8.5.3 Thickness Tests

Electrolytic thickness testing equipment or a chemical drop test can be used for determining chromium thickness. Microscopic, magnetic, and electrolytic methods are suitable for checking the nickel and copper thickness. Beta back-scatter or electrolytic equipment may be used to test the electroless deposits. X-ray florescence equipment may also be used to test electroless and electroplated deposit thicknesses.

8.5.4 Special Test

Expose the part in hot water, 71°–100°C (160°–212°F), for 2 min; then place it immediately into cold water, 4°–21°C (40°–70°F), for 2 min; then return it to the hot water. This is a fast thermal cycle shock test that may be useful as an "on-the-line" test.

8.5.5 C.S.A. Test

The Canadian Standards Association has developed the following thermal cycle test: ". . . an alternating dip test in water baths maintained at 79°C + 0°C − 2°C (175°F + 0°F − 2°F) and 21°C + 0°C (70°F + 2°F − 0°F). Parts should be totally immersed in each bath for 40 seconds—in the hot bath for 40 seconds followed by 40 seconds in the cold bath." Minimum time is allowed for the transfer from bath to bath.

The C.S.A. dip test with a 12-sec transfer time has proved to be a more severe test than the "standard" three-cycle thermal test of 82°C to 21°C to − 29°C (180°F to 70°F to −20°F), with 1-hr stabilization periods. The requirement of 450 cycles naturally means that a machine must be constructed for the test.

9

Automated Processing Equipment for the Electroplating of Plastic Substrates

9.1 REQUIREMENTS FOR PLATING ON PLASTICS

There are two types of plating on plastics provided today, decorative and functional. Decorative plating includes the electroless preplate cycle followed by copper, nickel, and chrome electroplate. Decorative electroplating requirements can be for either interior or exterior specifications. Functional requirements call for the chemical preparation of the substrate followed by electroless copper and electroless nickel depositions, primarily for EMI Shielding requirements. Either of the above processes can be easily automated.

9.2 CONSIDERATIONS FOR AUTOMATION OF PLATING ON PLASTIC PROCESSES

There are several considerations which must be explored before entering the world of electroplating automation. Not the least of these is answering the primary question, why automate? While there is no single answer to this question, most companies consider automation for two initial but extremely important reasons. First, to increase production throughput with fewer people, and secondly, to achieve more consistent results, therefore improving overall product quality with fewer rejects. Most all electroplaters of plastics would consider both of the above reasons as basic requirements for implementing the use of automated equipment.

Once the reasons for automation have been defined, the next consideration will revolve around choosing which type of automated equipment is best implemented. Because both decorative and functional finishing generally require rack-type processing, and, since we are addressing only two types of automated equipment in this chapter, namely the programmed hoist and the fixed-cycle return-type machine, the individual choice generally comes down to production requirements and versatility.

9.3 EQUIPMENT SELECTION

If the reason for automation is production driven, the return-type machine is generally, though not always, the equipment of choice. At production rates of sixty racks per hour or more, the RTM (return-type machine) is historically more cost effective. If versatility and/or flexibility is required or production rates are lower than sixty racks per hour, the programmed hoist system will meet the requirements for most plastic plating applications.

Typical production rates for a straight line one hoist system is 8 loads per hour, two hoists 14 loads per hour, three hoists 18 loads per hour and so on. Production rates for U-shaped lines and processes which require many steps as those for plating on plastics, the production rates are lower and additional hoists must be provided to reach required production levels.

9.4 AUTOMATED HOIST LINES

There are three primary styles of automated "hoist" lines manufactured by equipment suppliers today. They are: (1) the overhead style, (2) the cantilevered or side-arm style, and (3) the railriding style. A brief description with some general perceptions of each style follows:

9.4.1 Overhead Style

The overhead style is the oldest and most common type of programmed hoist in production today. It can be either a top riding or underhung style. It can be chain (rack and pinion) driven or friction driven. Lifting can be accomplished by either an end pick-up arrangement or a center column assembly. Travel speeds are variable up to 300 feet per minute. Lifting speeds are variable up to 300 feet per minute. Lifting speeds for work bars containing plastic plated parts are in the range of 65 feet per minute. These attributes are common for the other hoist styles as well.

Fig. 9.1 Overhead style

The superstructure required for support of the overhead hoist system is a floor mounted inverted "U" frame. Two primary advantages of the overhead hoist system include the lack of a weight restriction on its lifting capacity and the lifting height requirements. The width of the work bar in the contra direction of hoist travel also has fewer constraints.

Fig. 9.2 Cantilever style

9.4.2 Cantilever Style

The cantilever of "side-arm" hoist was introduced for use in the marketplace around 1965. It is supported from a floor mounted superstructure/tank base and can be either chain or friction driven. The lifting motion is accomplished from a lift carriage/boom assembly which extends over the process line from the main body of the hoist assembly. The lifting and travel motors are mounted to the side of the hoist and are less likely to be affected by the fumes from the process line. The primary disadvantage of the cantilever hoist is its restriction of a maximum 1000 pound lift capacity. Work bar width is also restricted to between 8 to 10 feet.

9.4.3 Rail Style

The rail style hoist has been used in Europe for over twenty-five years for all types of wet processing requirements and was most recently introduced into the US

Fig. 9.3A,B Rail style

market in 1985. The rail style hoist is a friction drive unit and offers the advantage of a floor mounted superstructure/tank base along with minimum constrains on lifting capacity weight restrictions. As with the cantilever style hoist, the rail style features motors mounted to the side of the hoist and away from the fuming from the process line. The rail style hoist also has minimum restrictions on the width of the work bar.

There are many additional options which can be provided with the programmed hoist. Among them, hoist mounted exhaust booths for either push or pull air. Totally enclosed systems are available. Hoists routinely incorporate central lubrication systems. Additionally, process line tanks are now provided with automatic covers controlled via the hoist control system.

9.4.4 Hoist Line Processing

Racks are suspended from a "work bar" and are transported via the hoist through the various process steps per the "timeway diagram", the timeway defined as the footprint for the hoist motions, or random process programs stored in the control system.

9.4.5 Hoist Positioning

There are three basic methods incorporated to provide on station hoist location. One such method would be the "dog and switch" using either a counting system or positive address system. Also, the rotary encoder, and lastly, the infrared light source are used.

9.5 AUTOMATED RETURN-TYPE MACHINES

As stated above, the return-type machine is chosen as the preferred automatic equipment when production rates exceed sixty racks per hour. There are two major styles of return type machines in production today: the elevator conveyor or side-arm, and the split-rail or pusher style machine. A brief description of each machine follows:

9.5.1 Elevator Conveyor

The elevator conveyor machine is a high volume rack processing machine used for all plating on plastics application. The machine may be either a single file (one rack per carrier arm) or double file (two racks per carrier arm) and may be equipped with automatic loading features or may be loaded manually, as required. Refer to Figure 9.5 for an automatic loading feature.

Fig. 9.4 Elevator Conveyor

The elevator conveyor is a forward indexing machine, i.e., lift, index forward, lower. The elevator lift mechanism as well as the forward drive unit may be powered by either hydraulic or electromechanical means. The forward drive is accomplished via a heat treated chain. The parts rack is suspended on a copper yolk attached to an aluminum carrier arm sized to carry DC current as required by the plating process. The carrier arm is attached to a lift carriage assembly mounted in a pair of channels. The carrier arm extends cantilevered over the tank line from the lift assembly thus keeping the central mechanism out of the process fumes.

9.5.2 Split-Rail Machine

The split rail style machine is also a high volume rack processing machine used for all types of wet processing requirements. As with the elevator conveyor,

Fig. 9.5 Elevator conveyor with automatic loading feature

the split-rail machine may be of either a single file (one rack per carrier arm) or double file (two racks per carrier arm) style and may be equipped with automatic loading features or may be loaded manually.

The split-rail machine is also a forward indexing machine, i.e. lift, index forward, lower. The lift and forward indexing for the split-rail machine is powered by a hydraulic unit. The rack is suspended on a copper work carrier which rides on the DC copper rail. The work carrier is "pushed" from one station to the next via a hydraulically operated cylinder as the machine cycles. The copper cathode rail is "split" at the entry and exit stations, lifting and lowering with the machine mechanism.

9.6 SYSTEM CONTROLS

Equally important in the selection process of the automatic equipment is the selection of the type of controller required for the automatic plating line.

Control systems have evolved over the years from simple electromechanical stepping switches, tenor drum or tape readers, to their present levels of sophistication incorporating the newest developments in personal computers (PC's) and programmable logic controllers (PLC's).

A key question which must be addressed pertaining to controls selection deals with processing requirements and the method of program selection. Will processing be random or via fixed timeways and will line process parameters be controlled or monitored via the hoist control system?

9.6.1 Computerized Control Systems for Plating on Plastics

The demands for higher productivity, better yields and lower total cost have increased the requirements for tighter control of both the preplate as well as electroplate processes associated with plated plastics industry. To fill this need, computerized control systems have evolved by offering complete automatic control of the plating system, monitoring and control of the plating line processes, and data logging and report generation.

The systems have succeeded at becoming extremely "user-friendly". The operator selects the appropriate command displayed on a screen. All information that needs to be entered is done through formatted screens using a flashing bar (cursor) which leads the operator through the entire operation. Some of the new systems similar to this one, actually prevent the operator from entering incorrect or conflicting information. Graphics supply the operator with up-to-date summary of the systems at a glance.

As shown in Figure 9.6, the basic functional blocks of this computerized control system consist of hoist control, process line control, and data logging.

Regardless of hoist type (overhead, cantilever, rail style, single or multiple units), the system controls vertical and horizontal hoist movements. The vertical speed of the hoist can be changed to minimize the effect of the work breaking through the surface of the bath. Extremely close tolerances can be held for horizontal positioning of the hoist. The rate of acceleration and deceleration and speed of the hoist can all be controlled during horizontal moves. This allows for faster moves and higher work throughput, while maintaining smooth work movement.

Using either a monochrome or color monitor, the hoist display indicates what process is being run, hoist mode, alarms, warnings, accumulated immersion times, the area of work in each tank and whether the line is being run by a time-immersion or sequence-mode program. For each hoist, the display indicates

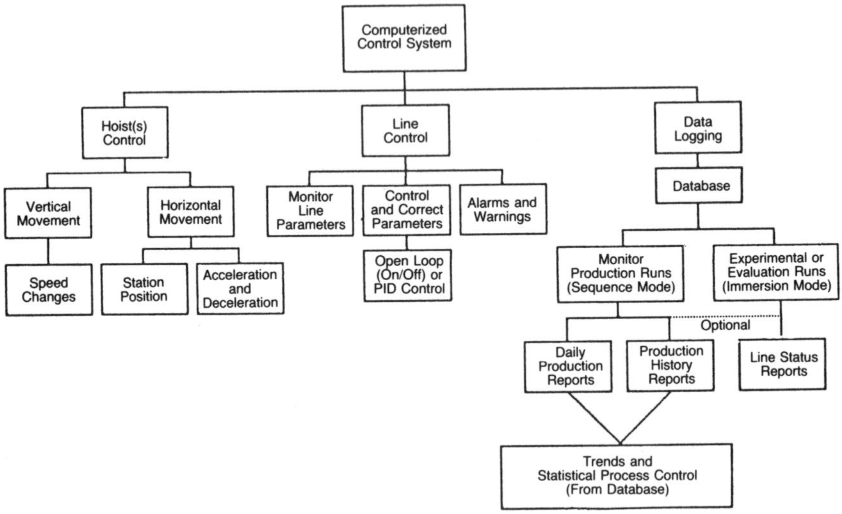

Fig. 9.6 Block diagram of computerized control system

the mode (automatic, manual or bypass) and, in the automatic mode, the process step being executed.

As shown in Table 1, the computerized control system monitors or controls all the process line parameters. Parameters can be controlled in either an open-loop (on/off) or closed-loop (PID) mode. Upper and lower set points as well as upper and lower warning levels are user selected. An external alarm trigger (digital input) can be assigned to any monitored parameter. The alarm trigger can be used to activate an audio or visual alarm or turn off either or both controls associated with the parameter. Typical parameters which can controlled and/or monitored include: temperature, conductivity, pH, bath replenishment, level, rectification, work bar tracking, flow rate, pressure, sprays, and ancillary equipment. The measurement display screen, shown in Figure 2, indicates the status of all communication lines and the status of all line parameters, showing station number, control status and the parameter's actual reading. Any alarm or warning, is indicated by flashing the out-of-spec parameter until it is corrected. Alarms and warnings plus any corrective action, must be acknowledged by the operator. These acknowledged are time-stamped if data logging is turned on. The limits of all parameters are automatically set to the values previously entered by the user in his process program, based on the process cycle file being run. The operator has the ability to edit the set points while the line is running.

Data logging is one of the most powerful features of the computer control system. Data is logged:

1. When work enters a station.

	Monitored	Controlled	Upper set point	Lower set point	Upper warning level	Lower warning level	On/Off control	PID control	Differential readings	External alarm trigger	Control status (On/Off)
Bath temperature	X	X	X	X	X	X	X	X		X	
Bath conductivity	X	X	X	X	X	X	X			X	
Bath pH balance	X	X	X	X	X	X	X			X	
Bath replenishment	X	X	X				X			X	
Fluid levels	X	X					X			X	
Flow rates	X				X	X				X	
Fluid pressures	X				X	X			X	X	
Spray rinses		X					X				X
Ancillaries		X					X			X	X
Rectifiers	X	X			X	X	X			X	

Table I. Line Parameters.

2. When work is removed from a station.
3. Periodically, while there is work in a station.
4. While any parameter is in a warning condition.
5. When there is a change in alarm or warning condition.

All acknowledgments are also logged. There is an option to log values periodically whether there is work present or not. Additionally, data logging can be turned on at any time, which allows monitoring while the plating system is in the manual or even the bypass mode.

The stored data is used to generate a variety of reports. These include production history reports daily production reports, and line status reports.

A database is generated by means of a data file which is created for each days production. This file is stored on the hard disc for up to 10 days and automatically purged to make room for current files. The actual storage time depends on the amount of data being stored. Since no predefined report formats can be expected to satisfy all requirements in all applications, report generators must be sufficiently flexible to permit customization by the process engineer.

The production history report is generated automatically and/or on demand. It consists of all data related to a particular work bar transferred through the plating line. Also indicated are warnings, alarms, deviations, start times, completion times, immersion times, and identification of the work load. This information gives the operator the ability to evaluate problems and relate them to causes.

The daily production report is generated automatically and/or on demand. This report is aimed at the production supervisor. Production statistics are supplied relating to a 24-hour period ending at the date and time of the report. The maximum number of work bars that could have been processed, in addition to the actual number processed, gives the line supervisor the ability to evaluate that day's productivity. The operational status, alarm history, and actual parts processed are neatly documented.

The line status report gives the chemist an accurate picture of the line status at a particular instant in time. This can be especially helpful during initial start-up of a new line or process.

9.6.2 Basic Controller

A basic control system incorporates the use of the PLC to control all machine functions and an operator interface console. Typical screens provided with this system include hoist status and program entry. Line parameters are controlled manually.

9.6.3 Random Processing

Random processing can be accomplished by using either a "priority" based system or a "reservation" based system. Both methods utilize either the PC or PLC. The "priority" system will allow the user to run multiple processes simultaneously but will not guarantee time in tank. The "reservation" system runs multiple processes simultaneously and guarantees time in tank.

While there are many different companies providing different styles of automated equipment, the basic types identified above remain the same. Each company offers their own approach to these basic designs as well as their own level of sophistication.

As you evaluate your automation requirements, either new or replacement, remember your equipment suppliers are professionals and experts in the wet processing automation field and will be of invaluable assistance to you.

Utilize this expertise and your automation project will be completed both economically and successfully.

10

Test Procedures and Standards for Electroplated Plastics

10.1 TESTING PROCEDURES AND STANDARDS

The test procedures and standards, as covered in this section, are very important in comparative studies of various combinations of metal plating on plastic substrate and evaluation of various production methods.

The words *comparative studies* are emphasized, since in any industry the guide to true effective life of a product is obtained through a cumulative history of service performances.

These tests are presented to obtain uniformity in test procedure, for inclusion in specification requirements, and to improve the accuracy and value of comparative testing.

These tests and standards will be subject to periodic review and revision by the Industry Standards Committee of the American Society of Electroplated Plastics, Inc.

10.2 STANDARD MOLD DESIGN FOR THE ASEP STANDARD TEST PANEL (ASEP STD-100)

10.2.1 Scope

1. The purpose of this Standard is to provide a standard mold design for the ASEP Standard Test Panel.

2. This Standard eliminates the mold design variable from test work being performed with the ASEP Standard Test Panel.

10.2.2 Design

The mold specifications and drawings are shown in Figure 10.1 and 10.2.[1]

10.3 RECOMMENDED PRACTICE FOR EVALUATION OF THE APPEARANCE OF PLATED PLASTIC SURFACES (ASEP STD-101; ASTM B-532)

10.3.1 Scope

1. This recommended practice offers information on the evaluation of the appearance of electroplated plastic surfaces. This is a suggested guide only. In any specific case, there should be a prior agreement between the purchaser and manufacturer regarding what constitutes an acceptable surface as far as appearance is concerned.

2. The appearance of surfaces will be separately evaluated, depending on the prominence of the surface under conditions of ordinary use.

10.3.2 Classification of Surfaces

1. Surface A: The surface that appears most prominent to the observer or most likely to be noticed at first glance.

2. Surface B: The surface (usually adjacent to Surface A) which is of relatively less prominence to the observer.

3. Surface C: The surface not readily seen by the observer except at an angle or by special observation.

4. Surface D: The surface not visible to the observer under conditions of normal use.

10.3.3 Method of Evaluation

1. The part under test shall be evaluated in the same relative position as it would normally be viewed under usual conditions of use.

[1]To get a full-size drawing, contact ASEP headquarters.

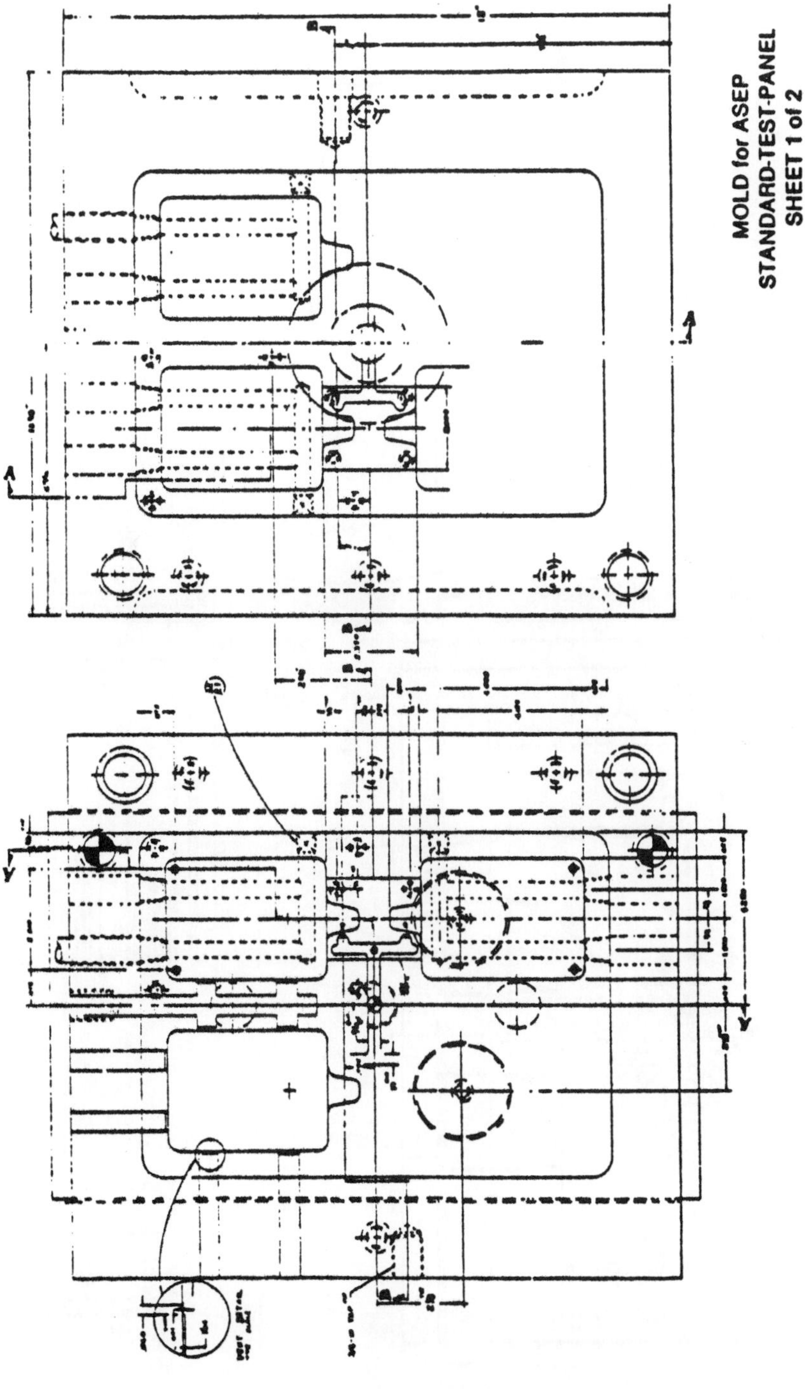

Figure 10.1

Figure 10.1

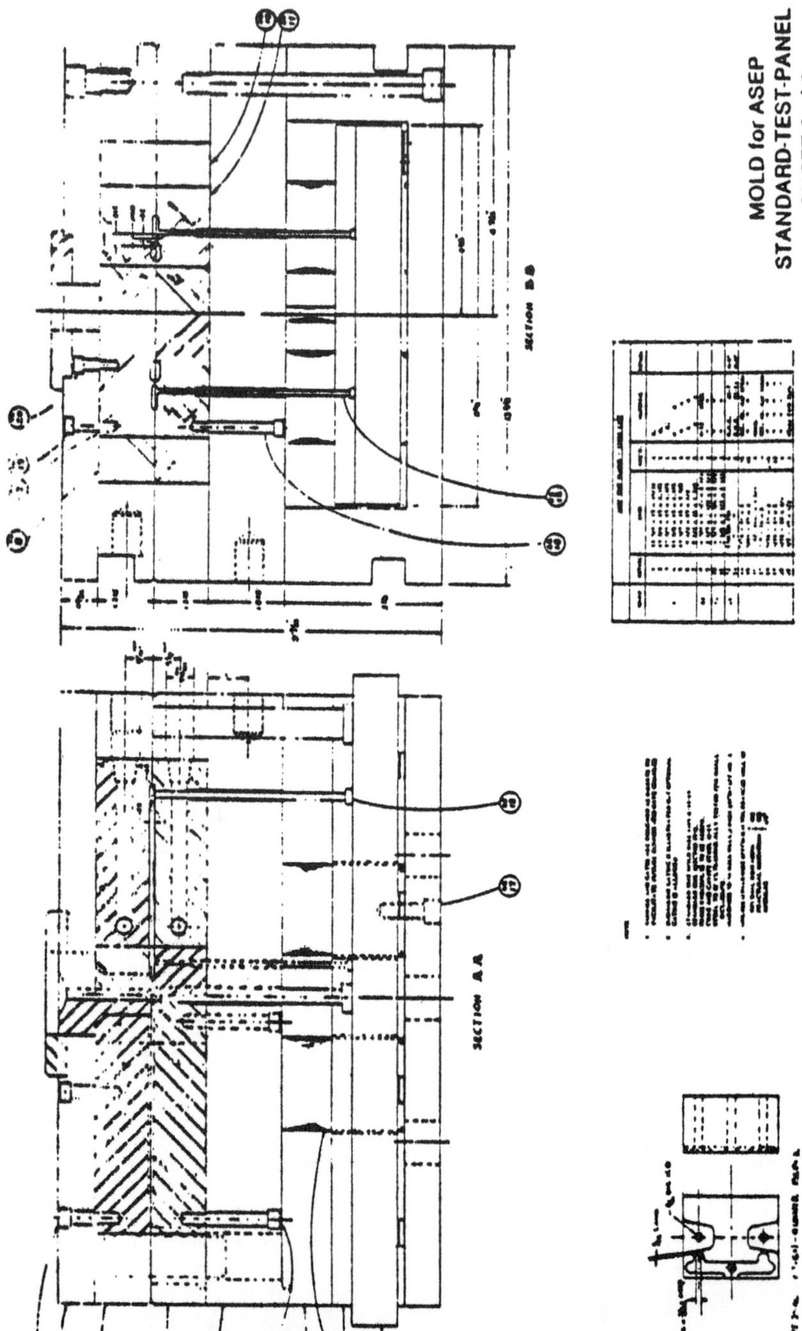
MOLD for ASEP
STANDARD-TEST-PANEL
SHEET 2 of 2
SECTION B-B
SECTION A-A

Figure 10.2

TABLE 10.1 Evaluation of Appearance of Plated-Plastic Surfaces

Defect	Surface			
	A	B	C	D
Cold Shot (M)	No	No	Some	Yes
Pits (M+P)	No	No	Some	Yes
Sink Marks (M)	No[2]	No[2]	Yes	Yes
Ejection Marks (M)	No	No	Yes	Yes
Blisters (P)	No	No	No	No[3]
Peeling (M+P)	No	No	No	No[3]
Roughness (M+P)	No	No	Some	Yes
Staining[1] (M+P)	No	No	Some	Yes
Burning (M+P)	No	No	Some	Some
Visible Intermediate Deposits (P)	No	No	No	Some[4]
Scratches, Nicks (M+P)	No	Some	Some	Yes
Parting Line (M)	No[2]	Some	Yes	Yes
Weld Line (M)	No[2]	Some	Yes	Yes
Gate Mark (M)	No[2]	Some	Yes	Yes
Flash (M)	No	No	No	No
Rack Mark (P)	No	No	No	Yes
Voids (M+P)	No	No	Some	Yes
Splay (M)	No	No	Some	Yes
Stardusting (P)	No	No	Some	Yes
Off Color (P)	No	No	Some	Yes

(M) These defects arise primarily in the plastic molding operation.

(P) These defects arise primarily in the plating operation.

(M+P) These defects may arise in either the molding or the plating operation.

[1]Not removable by wiping or cleaning.

[2]Existence of these defects should be noted and resolved prior to tool manufacture. A change of design may be required.

[3]Blisters or peeling in the gate area or in the plating-rack contact may be acceptable if the given area is covered or otherwise hidden after assembly.

[4]May occur at rack contacts or low-current-density areas.

2. After proper positioning, the part under test will be evaluated at a distance of not less than 24 in nor more than 36 in from the observer, using diffused-daylight-type fluorescent lighting of about 150 foot-candles.

3. The observer shall have 20/20 vision, corrected with eyeglasses if necessary.

10.3.4 Appearance Guide

1. Table 10.1 is a recommended practice for the evaluation of appearance, subject to the limitations of section 10.3.1.1.

2. In Table 10.1, the term *some* assumes that the defects cited are not outstanding, too numerous, or located too closely together. The term *no* indicates that the defect cited is not normally acceptable. The term *yes* indicates that the defect cited is normally acceptable.

10.4 STANDARDS AND GUIDELINES FOR THE PERFORMANCE OF DECORATIVE ELECTROPLATED COATINGS OF COPPER-NICKEL-CHROMIUM ON PLASTIC (ASEP STD-102)[2]

10.4.1 Scope

These Standards and Guidelines cover requirements for several types and grades of electrodeposited copper-nickel-chromium coatings on plastic substrates for applications where both appearance and durability of the coating are important. Four grades of coatings are provided to correlate with the service conditions under which each is expected to provide satisfactory performance; namely, very severe, moderate, and mild. Definitions and typical examples of these service conditions have been established by ASTM and are provided in section 10.4.10.

10.4.2 Manufacture

1. Defects in the substrate, such as scratches, pores, die marks, and cracks, may adversely affect the appearance and the performance of coatings applied to the surface despite the observance of the best electroplating practice. Accordingly, the electroplater's responsibility for defects in the coating resulting from such conditions shall be waived, except when he is also in the position of prime contractor supplying electroplated parts.

Note 1: Evaluation of plastic surfaces for electroplating is discussed in more detail in the ASEP Standards and Guidelines in Chapter 1.

2. Proper preparatory procedures are essential in order to assure satisfactory adhesion and corrosion performance of the coatings. With plastic substrates, corrosion of the plastic is practically nonexistent; therefore, durability of the electrodeposited coating is extremely important in order to prevent penetration of the coating, which may lead to corrosion or blistering of the coating itself.

Note 2: Refer to the sections of the ASEP Standards and Guidelines discussing Handling of Molded Parts for Plating found in section 4.4.8.

[2]The information contained in this section was developed in cooperation with ASTM members and has been incorporated in ASTM Standard B-604-75.

10.4.3 Significant Surfaces

Significant surfaces are defined as those normally visible (directly or by reflection) that are essential to the appearance or serviceability of the article when assembled in normal position or that can be the source of corrosion products that deface visible surfaces on the assembled article. When necessary, the significant surfaces shall be the subject of agreement between purchaser and manufacturer and shall be indicated on the drawings of the parts or by the provision of suitably marked samples.

Note 3: When significant surfaces are involved on which the specified thickness of deposit cannot readily be controlled, such as threads, holes, deep recesses, bases of angles, and similar areas, the purchaser and the manufacturer should recognize the necessity for either thicker deposits on the more accessible surfaces or for special racking. Special racks may involve the use of conforming, auxiliary, or bipolar electrodes or shields (conducting or nonconducting).

10.4.4 Appearance

1. The significant surfaces of the plated article shall be free from clearly visible plating defects, such as blisters, pits, roughness, cracks, or unplated areas and shall not be stained or discolored. On articles where a visible contact mark is unavoidable, its position shall be the subject of agreement between the manufacturer and the purchaser. (See Note 1.)

2. The plated article shall be clean and free from damage. The purchaser shall specify the appearance required; for example, bright, dull, or satin. Alternatively, samples showing the required finish or range of finish shall be supplied or approved by the purchaser.

10.4.5 Manner of Specifying Requirements

1. Coating Classification Number or Service Condition Number. When ordering articles to be plated in accordance with this standard, the purchaser shall state either the service condition number denoting the severity of the conditions it is required to withstand (See 10.4.5.2) or the classification number of the particular coating required (See 10.4.5.4). If the service condition number is quoted and not the classification number, the manufacturer is free to supply any of the classes of coating corresponding to the service condition number, but he shall inform the purchaser of the classification number of the coating supplied, when requested.

2. Service Condition Number. The service condition number indicates the severity of the service conditions in accordance with the following scale:

SC 4—very severe service
SC 3—severe service
SC 2—moderate service
SC 1—mild service

Typical service conditions for which the various service condition numbers are appropriate are given in section 10.4.10.

3. Coating Classification Number. The coating classification number is comprised of the following:

The symbol for the substrate, PL, indicating that the substrate is plastic.

The chemical symbol for copper (Cu).

A number indicating the minimum thickness of the copper coating, in micrometers (Note 4).

A lower-case letter designating the type of copper deposit.

The chemical symbol for nickel (Ni).

A number giving the minimum thickness of the nickel, in micrometers (Note 6).

A lower-case letter designating the type of nickel deposit.

The chemical symbol for chromium (Cr).

A lower-case letter or letters designating both the type of chromium and its minimum thickness in micrometers.

Note 4: The approximate equivalent thickness in mils is included in Table 10.2 (1 μm = approximately 0.04 mil; 1 mil = .001 in = 25μm).

Type of Copper. The type of copper (Note 6) is designated by the symbol *a* for ductile copper deposits prepared from acid baths containing additives that promote leveling of the copper deposit and that pass the bend test method given in section 10.4.11.

Note 5: It is not yet known whether copper deposited from cyanide, pyrophosphate, or other solutions would provide adequate corrosion protection and durability. This specification recommends the use of copper deposits prepared from acid baths now available from plating suppliers.

Type of Nickel. The type of nickel (Note 6) is designated by one of the following symbols: *b* for nickel deposited in the fully bright condition, or *d* for a multilayer coating consisting of two or three layers, of which the bottom layer contains less than 0.005% sulfur (Note 9) and that passes the bend test given in section 10.4.11. The top layer generally contains more than 0.04% sulfur. In double-layer coatings, the thickness of the bottom layer shall not be less than 75%

TABLE 10.2 Copper-Nickel-Chromium Coatings on Plastic*

Service Condition Number	Classification Number	Equivalent Nickel Thickness μm	Equivalent Nickel Thickness mils (approximate)
SC 4	PL/Cu15a Ni30d Crr	30	1.2
	PL/Cu15a Ni25d Crmc	25	1.0
	PL/Cu15a Ni25d Crmp	25	1.0
SC 3	PL/Cu15a Ni25d Crr	25	1.0
	PL/Cu15a Ni20d Crme	20	0.8
	PL/Cu15a Ni20d Crmp	20	0.8
SC 2	PL/Cu15a Ni15b Crr	15	0.6
	PL/Cu15a Ni10b Crmc	10	0.4
	PL/Cu15a Ni10b Crmp	10	0.4
SC 1	PL/Cu15a Ni7b Crr	7	0.3

*The minimum copper thickness may be greater in some applications to meet thermal cycling and other requirements.

of the total nickel thickness. In triple-layer coatings, the bottom layer shall not be less than 50% of the total nickel thickness. The thickness of the top layer in either case shall not be less than 10% of the total nickel thickness. If there are three layers, the intermediate layer shall contain more sulfur than the top layer and shall not exceed 10% of the total nickel thickness.

Note 6: The type of copper and of nickel will usually be identifiable by microscopic examination of a polished and etched section of an article, prepared in accordance with ASTM Method B-487.

Note 7: The sulfur contents are specified in order to indicate the type of nickel. Sulfur contents can be determined by chemical analyses on specifically prepared test specimens. No simple, reliable method exists for determining the sulfur content of a nickel deposit once it is applied to an article.

Thickness of Copper and Nickel Deposits. The number following the chemical symbols Cu and Ni indicates the minimum deposit thickness, in micrometers (Note 4), measured in accordance with ASTM test methods at points on the significant surface. The following test methods[3] may be used, as appropriate:

[3]Copies of ASTM Standards are available from ASTM Headquarters, 1916 Race Street, Philadelphia, PA 19103.

Method B-487, if coatings are thicker than 10μm (0.4 mil).

Method B-504, if the coating is between 0.75 and 50μm (0.03 and 2 mils).

Method B-530, when the thickness of the nickel in the multilayer deposit is to be determined.

Method B-567, when the composite coating thickness is to be determined.

Other methods can be used if the uncertainty of the measurements is shown to be less than 10%.

Type of Chromium and Deposit Thickness. The thickness of the chromium deposit shall be measured at points on the significant surface by the test methods given in either Methods B-504 or B-556.

The type of chromium and thickness of deposit are both designated by the following symbols placed after the chemical symbol Cr (numerals are not used in this case to specify thickness as in the case of nickel):

r for regular (that is, conventional) chromium, having a minimum thickness of 0.25μm (0.01 mil), except that for Service Condition 1 the minimum thickness may be reduced to 0.13μm (0.005 mil).

mc for microcracked chromium, having more than 300 cracks/cm (750 cracks/in) in any direction over the entire significant surface, and a minimum thickness of 0.25μm (0.01 mil).

mp for microporous chromium, having a minimum thickness of 0.25μm (0.01 mil) and containing a minimum of 10,000 pores/cm^2 (64,500/in^2) invisible to the unaided eye.

Example of a Complete Classification Number. A coating comprising 15μm (0.6 mil) of minimum copper, 15μm (0.6 mil) of minimum bright nickel, plus 0.25μm (0.01 mil) of minimum microcracked chromium has the classification number PL/Cu 15a Nil5b Crmc.

4. Coatings Appropriate to Each Service Condition Number. Table 10.2 shows the coating classification numbers appropriate for each service condition number.

10.4.6 Thermal Cycle Requirements

1. Plated articles shall be subjected to four cycles of the thermal cycle test as outlined in ASEP Standards and Guidelines. The service condition number specified (or the service condition number for the coating classification number; see Table 10.2) shall correspond to the service condition number for determining the temperature extremes.

TABLE 10.3 Corrosion Tests Appropriate for Each Service Condition Number

Service Condition Number	Corrosion Test Duration (CASS ASTM B-368)
SC 4	Three 16-hour cycles
SC 3	Two 16-hour cycles
SC 2	8 hours
SC 1	. . .

2. After having been subjected to four cycles of the appropriate thermal cycle test, the coated article shall show no visible defects, such as cracking, blistering, peeling, sink marks, or distortion.

Note 8: In many cases, there is no direct relation between the results of thermal cycle testing and performance in service, because it is not always possible to predict and control the thermal exposure of the coated article in service or during storage. Therefore, the results of thermal cycling should be used to monitor the quality of plated-plastic articles and not as a direct guide to performance in service.

3. Articles plated and thermal-cycle-tested in accordance with Service Conditions 3 and 4 shall be cut with a fine-toothed hacksaw. There shall be no evidence of separation of the coating from the plastic surface or separation between layers of the coating along the exposed cross section of the plated article.

Note 9: ASEP provides a procedure for measurement of the peel strength (adhesion) of metal-plated plastics using standard specimens. Since there is no direct correlation between results obtained on standard specimens and actual molded parts, the method does not assure good adhesion on molded parts but is useful to determine that the processing solutions are capable of giving acceptable results.

10.4.7 Corrosion Resistance or Corrosion Protection of Coatings

1. Coated articles shall be subjected to the corrosion test for a period of time that is appropriate for the particular service condition number, as shown in Table 10.3.

Note 10: In many instances, there is no direct relation between the results of an accelerated corrosion test and the resistance to corrosion in other media, because several factors that influence the progress of corrosion, such as the formation of protective films, vary greatly with the conditions encountered. The results obtained in the test should not, therefore, be regarded as a direct guide to the corrosion resistance of the tested materials in all environments where these materials may be used. Also, performance of different materials in the test cannot always be taken as a direct guide to the relative corrosion resistance of these materials in service.

2. After subjecting the article to the treatment described in the relevant test method, it shall be examined for evidence of corrosion penetration to the substrate or the copper layer and blistering of the coating. Any evidence of copper corrosion, blistering of the coating, or substrate exposure shall be cause for rejection.

Note 11: It is to be understood that occasional widely scattered, small corrosion defects may be observed after the testing period. In general, acceptable resistance shall mean that such defects are not, when viewed critically, significantly defacing or otherwise deleterious to the function of the plated part.

3. Surface deterioration of the coating itself is expected to occur during the testing of some types of coatings.

10.4.8 Combined Thermal Cycle and CASS Testing[4]

The thermal cycle and CASS tests may be combined by using the same samples in each test and by running the tests in sequence as described in this section.

Note 12: Although agreement is not universal, many feel that combined thermal cycle-CASS testing is more severe and, hence, better controls the quality of electroplated-plastic parts than individual tests run on separate specimens.

1. Expose the parts to one 16-h cycle of CASS testing.
2. As soon as possible, expose parts electroplated according to the requirements of Service Condition 4 for 1 hr at ± 1.5°C. For Service Condition 3, expose the parts for 1 hr at ± 1.5°C.
3. Remove the parts and expose them for 15 min at 25°C ± 5°C.
4. Expose the parts for 1 hr at 85°C ± 1.5°C.
5. Expose the parts for 15 min at 25°C ± 5°C.

These five steps represent one cycle of the combined thermal cycle-CASS testing. For parts electroplated to either SC 4 or SC 3, repeat three additional

[4]The use of combined thermal cycle–CASS testing obviates the need to conduct the individual tests described in 10.4.6 and 10.4.7.

times. Examine parts for signs of adhesion loss, blistering, crazing, lifting, corrosion, and other defects after each cycle.

10.4.9 Sampling

Since test methods may be destructive, thereby rendering 100% inspection impracticable, it is recommended that the purchaser select suitable sampling plans for the acceptance testing of lots of coated items. General information on sampling procedures is given in ASTM Recommended Practice E-105 and in ASTM Recommended Practice E-122. Standard sampling plans have been published by several sources. In order that the manufacturer knows the quality he is expected to meet, the plans selected should be made part of the purchase contract.

10.4.10 Definitions and Examples of Service Conditions for Which the Various Service Condition Numbers Are Appropriate

SC 4 (Very Severe). Service conditions that include likely damage from denting, scratching, and abrasive wear in addition to exposure to corrosive environments and temperature extremes; for example, conditions encountered by exterior components of automobiles and by boat fittings in salt-water service.

SC 3 (Severe). Exposure that is likely to include occasional or frequent wetting by rain or dew or strong cleaners and saline solutions and temperature extremes; for example, conditions encountered by porch and lawn furniture, bicycle and perambulator parts, and hospital furniture and fixtures.

SC 2 (Moderate). Indoor exposure in places where condensation of moisture and temperature extremes may occur; for example, in kitchens and bathrooms.

SC 1 (Mild). Indoor exposure in normally warm, dry atmospheres with coating subject to minimum wear or abrasion.

10.4.11 Ductility Bend Test (See ASTM Recommended Practice B-489)

Note 13: This is used to test compliance of the type of copper or nickel deposit as outlined in section 10.4.5.3 and should not be used to assess the acceptability of a plated article.

1. Preparation of Test Piece. Prepare a plated test strip, 150 mm long, 10 mm wide, and 1 mm thick (approximately 6 by 0.4 by 0.04 in) by the following method:

> Prepare a sheet of steel. (Use a sheet sufficiently large to allow the test strip to be cut from it after trimming off a border at least 25 mm (1 in) wide all around.) Plate one side of the sheet with the deposit to be tested to a thickness

of 25μm (1 mil) under the same conditions and in the same bath as the corresponding articles.

Cut the test strip from the plated sheet with a flat shear. Round or chamfer the longer edges of the test strip, at least on the plated side, by careful filing.

2. Procedure. Bend the test strip with the plated side in tension, by steadily applied pressure, through 180° over a mandrel of 12-mm (0.45-in) diameter (Note 14) until the two ends of the test strip are parallel. Ensure that contact between the test strip and the mandrel is maintained during bending.

Note 14: The mandrel diameter of 12 mm (0.45 in) will produce approximately 8% elongation in the coating if there are no cracks passing completely across the convex surface.

3. Assessment. The plating complies with the minimum requirement provided that after testing there are no cracks passing completely across the convex surface. Small cracks at the edges do not signify failure.

10.5 STANDARD METHOD OF TEST FOR THE PEEL STRENGTH OF METAL PLASTICS (ASEP TP-200: ASTM B-533)

10.5.1 Scope

This method of test measures the peel strength between a metal plating and its plastic substrate when tested on standard-sized specimens and with specified conditioning of the panels, thickness of plating and substrate, rate of pull, and molding conditions.

Note 1: The so-called adhesion of a metal electrodeposit to a plastic substrate, as measured by the peel test, is not a measure of the attachment of that film to the substrate but rather a measure of the strength of the weakest layer in the metal-substrate combination. The peel test, as developed by Jacquet, gives a quantitative value of this combination.

10.5.2 Method

A tensometer is used to measure the tensile load, acting at 90° ± 5° to the substrate surface, which will peel a 1-in width of metal plating from its substrate at a rate of 1.0 ± 0.1 in/min. The film tested shall be that of the first substantial electroplated film only.

Note 2: Participants of an interlaboratory study shall agree beforehand what the first substantial film shall be.

Note 3: All test data will appear in English units rather than metric units. Direct conversion between English and metric units cannot be made.

10.5.3 Apparatus

1. For research activities, the universal tensile testing machine or equivalent *should* be used.

Note 4: The Instron machine has been found satisfactory. Available from Instron Engineering Corporation, 2505 Washington Street, Canton, MA 02021.

Note 5: In using the universal tensile testing equipment, some type of parallelogram should be attached to the upper fixed head so as to maintain a 90° pull angle on the strip. An example is shown in Figure 10.3.

2. For quality control activities, an indicating-force instrument may be used.

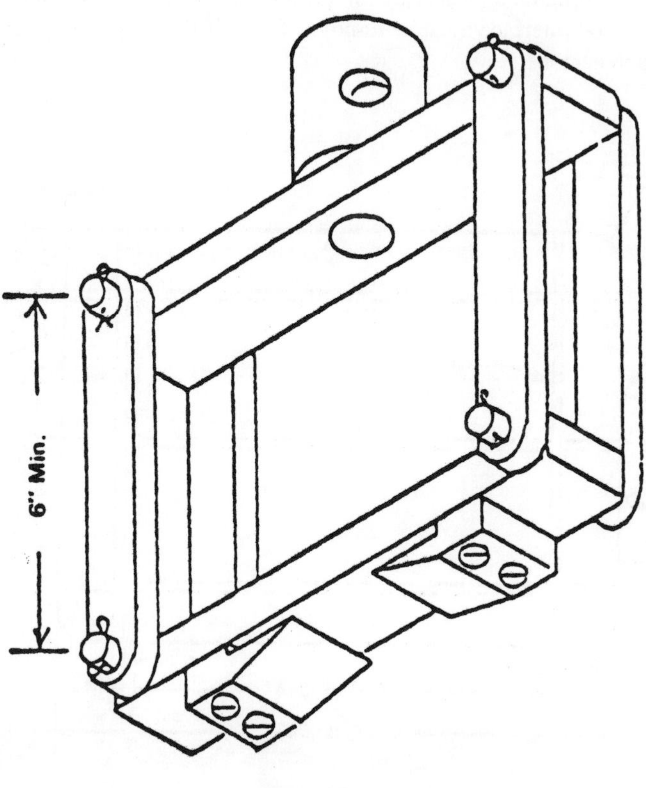

Figure 10.3

Note 6: The "Platron," Model 1, or equivalent, has been found satisfactory. Available from Plateq Corporation, 250 Sackett Point Road, North Haven, CT 06478.

10.5.4 Number of Tests

At least two plaques shall be tested in the unconditioned state and at least two in the conditioned state. (See section 10.5.6.)

10.5.5 Test Specimen Preparation

1. Use plaque as shown in Figure 10.4.
2. Plate the required number of plaques, using the plating or molding condition under evaluation, to a thickness of .0015 ± .0002 in.

Note 7: Plating thickness may be measured by ball micrometer after the peel test is completed. The film of plastic substrate usually adhering to the substrate-metal interface must be removed first. (If the plastic is ABS, it may be dissolved in acetone.)

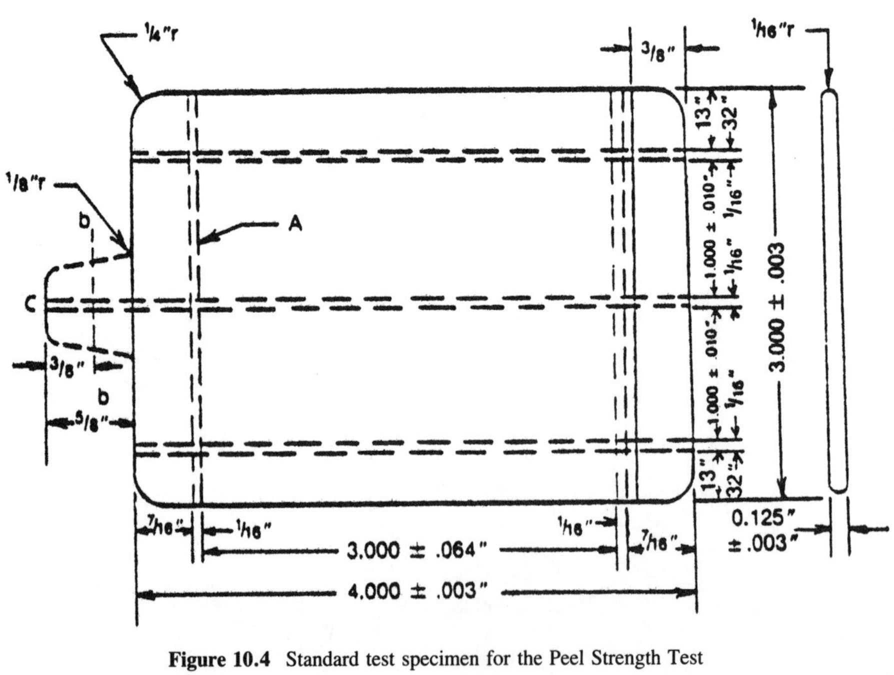

Figure 10.4 Standard test specimen for the Peel Strength Test

3. The conditioned plaque shall be heat-treated at this point. (See section 10.5.6.)

4. Cut through the metal layer to form the 1-in strips as shown in Figure 10.4.

Note 8: This may be done by one of the following methods:

Use any standard milling machine with a sharp slitting saw rotating into the plated film at 60 S.F.M. and a feed of 1 ft/min. A Union HSG-#14-0.064 slitting saw of Union Twist Drill Company, Athol, MA 01331, has been found satisfactory.

Use a slitting device based on Figure 10.5.

Use a sharp knife or razor blade and a straight edge. The "Specimen Scriber" of Plateq Corporation, or equivalent, (see Note 6) has been found satisfactory.

5. Starting at saw cut A, peel back the metal layer for a distance of about ⅝ in by crosscutting under the metal film. This is best done with a sharp knife or sharp 1-in chisel.

Note 9: An X-Acto knife, with #16 blade, has been found satisfactory. Heat should never be used to effect separation.

6. The attachment of fiberglass-reinforced self-adhering tape to the loosened tag is optional. However, if it is used, it shall not exceed beyond the ⅝-in tag length.

10.5.6 Test Specimen Conditioning

1. The plaque shall be plated and tested between 30 to 60 min after plating. The result of this test shall be reported as the "unconditioned" result.

2. Within 1 hr after plating, the plaque shall be placed in a circulating oven at 85° ± 1.5°C for 1 hr, cooled to 25° ± 5°C, prepared, and tested. This test shall be reported as the "conditioned" result.

Note 10: Modified polyphenylene oxide must be aged 12 to 16 hr to give a stable reading.

Note 11: Adhesion values change with time after the final plating operation. They may continue to change with time even after the conditioning treatment.

10.5.7 Test Procedure

1. Calibrate the transometer by use of known dead weights.

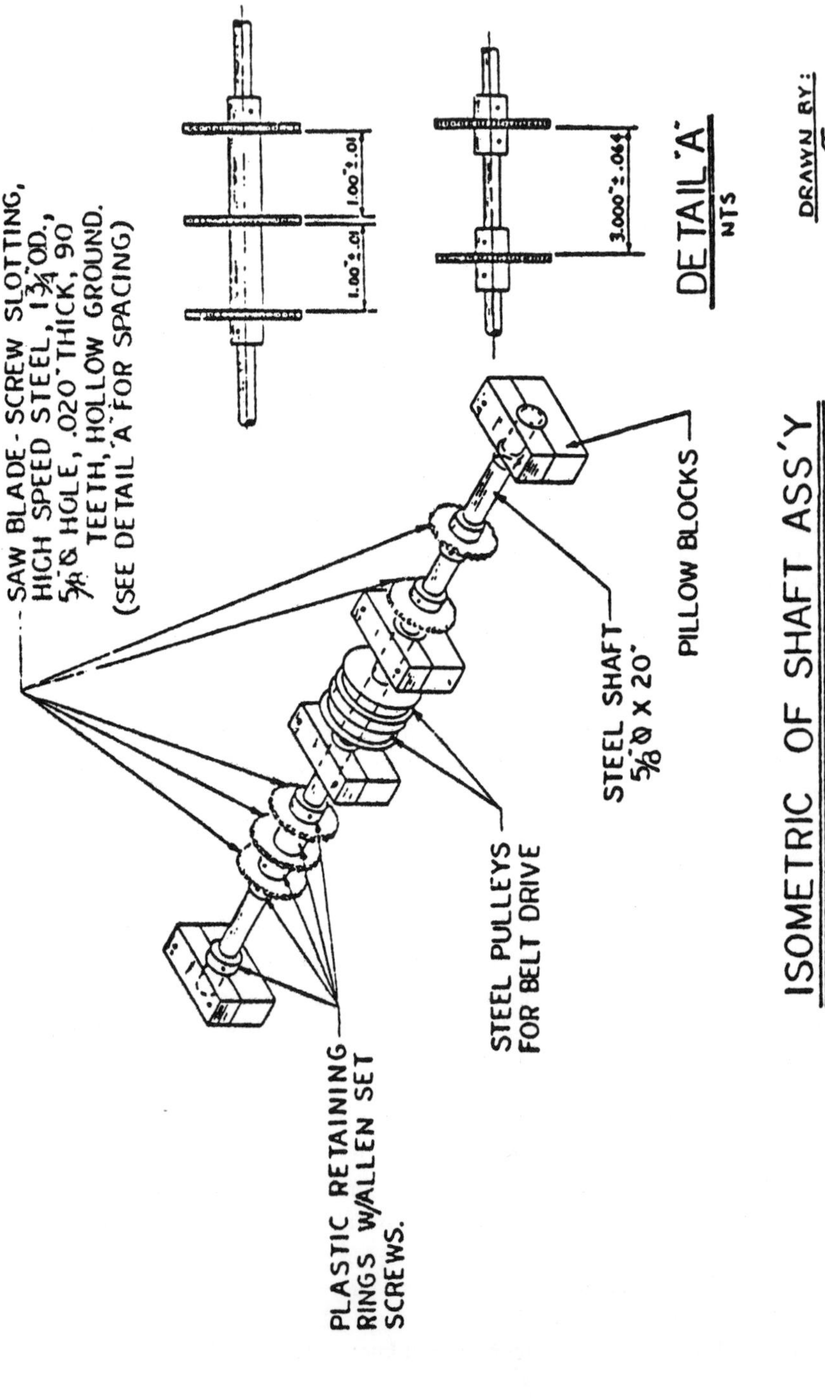

Figure 10.5 Table-slitting device

2. Place the plaque in the appropriate test instrument and fasten the tag to the movable clip.

3. Peel the plated film from the substrate at a rate of 1.0 ± 0.1 in/min at 90° ± 5° to the surface of the panel.

4. On the quality-control force-indicator instrument, record all the peak values obtained. On the universal tensile tester, record indicated values every 10 sec, if not equipped with a recording attachment, or identify panel being tested on chart strip, if equipped with recording device.

5. Repeat with the second strip on the plaque.

6. If the metal strip tears sideways, so that less than a 1-in width is being peeled, the test shall be discontinued, but readings up to this point may be used, provided that at least one 1-in length has been pulled properly.

10.5.8 Calculations

1. On the spring scale type of instrument, use the arithmetical average of the recorded high peak readings.

2. On the universal tensile tester type of instrument, use the arithmetical average of the recorded readings or draw a median line on the chart and use this value.

3. The averages are to be determined separately for each strip tested.

10.5.9 Report

The test report shall include the following:

1. Generic type, manufacturer's grade, lot number of the substrate, and molding conditions.
2. Plating batch number and plating processes used.
3. Average peel strength in pounds per inch of width, the highest and lowest value obtained, and statement as to whether panel is conditioned or unconditioned.
4. The number of tests made and the number of panels tested.
5. Type of first substantial electroplated film.
6. Thickness of the electroplated film tested, if other than .0015-in.

10.5.10 A Note on Results

In studies involving substrate variables, the plating conditions are held constant. In studies involving variables in the plating process, the conditions for producing the substrate are held constant. In the latter case, the recommendations of the material supplier should be followed. Also, it must be remembered that there is no direct correlation between results obtained on this "standard panel" and actual molded parts. Satisfactory results obtained in the plating of this standard panel mean that all the processing solutions are functioning correctly and are being used properly. This is true since the molding conditions affect the peel strength greatly even though all plating conditions are constant. Variations in peel strength will be found on different areas of an actual molded part. It is, therefore, advisable to mold rather large numbers of test specimens at one time to minimize batch variations.

10.6 DYNAMIC TEST OF PLATE ADHESION BY THERMAL CYCLING (ASEP TP-201)

10.6.1 Scope

1. This method describes the apparatus and procedures for evaluating plate adhesion over the entire surface of electroplated plastic parts and may be used for any plateable plastic.

2. This method is recommended for use in evaluating and controlling the entire manufacturing process, including molding, preplate, and electroplating processes.

3. This method tests the ability of the plate adhesion to withstand the dynamic forces at work as a result of the different thermal coefficients of expansion of the plastic and the metal electroplate when temperatures approach the heat-distortion point of the plastic.

4. This test, properly performed and with regular and adequate A.S.L. sampling, will guard against poor adhesion or nonadhesion of the plating. It is necessary to test 100% of the surface to identify adhesion problems in areas of excessive gate blush and molding stress and in cases of poor etching or poor part design. Experience has proven that parts with poor adhesion or nonadhesion that will not pass the dynamic thermal cycle test will have a higher than average probability of failure of the electroplating in its intended use.

10.6.2 Apparatus

The apparatus shall consist of a circulating-air heating chamber and cooling chamber sufficiently powered, insulated, and controlled so as to maintain closely the set temperature. The controller and recorder used for chamber control, calibration, and records must be accurate to 0.6°C (1°F). All points within the working area of the test chamber must remain within 2.8°C (5°F) of the set temperature when tested by the following procedures.

1. Install eight 24-gage bare thermocouples 2.5 cm (1 in) from each corner of the working area of the test chamber, plus one in the center and one at the sensing point of the chamber's controller, for a total of ten thermocouples.
2. Calibrate each thermocouple in boiling water and ice water after test setup is complete. (Also, note ASTM E-220, "Comparison Calibration of Thermocouples.)
3. Load chamber with maximum quantity and type of parts to be tested.
4. Determine boundaries of control: location within chamber, maximum loading, size of sample.

10.6.3 Test Specimen

The samples (test specimens) shall be drawn at random from the quantity under evaluation. The samples shall be introduced into the chamber unmounted.

10.6.4 Conditioning

1. Since the time between plating and thermal cycling may have varying effects on thermal cycle results, depending on the plastic being plated, tests should be run in order to determine the effects of conditioning time.
2. A standard conditioning time should then be specified for a given plastic so that comparative analyses will be meaningful.

10.6.5 Procedure

1. The thermal cycling routine to be used includes a high limit of 5.6°C (10°F) less than the heat-distortion point of the plastic being tested. The plastic manufacturer's stated unannealed heat-deflection temperature at 1820 kPa (264 psi) according to ASTM D-648 shall be used as the basis for determining the high limit.
2. The low limit shall be −40°C (−40°F) ± 1.5°C.
3. Each thermal cycle will consist of placing the samples directly into a chamber at the high limit and performing the following:

    1. One hour soak at the high limit.
    2. One-half hour at 25°C ± 5°C.
    3. One hour at the low limit.
    4. One-half hour at 25°C ± 5°C.
    5. Part examination and recording of defect type and location.

10.6.6 Qualification Testing

Qualification testing may be initially performed to define the failure modes and percentage of failures as a function of cycle number. Ten thermal cycles should be performed (or 50% cumulative failures) so as properly to characterize the results of a system consisting of substrate, part design, mold design, molding, and plating. Sample size must be sufficient to insure adequate confidence in the estimates obtained.

10.6.7 Acceptance Testing

Acceptance testing consists of drawing a sample from each production lot or shipment and exposing this sample to three thermal cycles. The sample size and acceptance number shall be determined from the specified AQL and inspection level or from agreed-upon, defined sampling risks. A prerequisite to acceptance testing is a classification of defects, defining quality levels for defect types and location.

10.6.8 Compilation and Evaluation of Results

Parts must be carefully inspected after each complete cycle for blisters, cracks, peeling, and other defects that were not in evidence prior to thermal cycling. The severity and location of each defect should be noted, as well as the number of thermal cycles completed at the time of the inspection. Data (failures) will then be categorized in such a manner that significant characteristics can be independently evaluated. These categories may be composed of defect types, sizes, or locations, depending on the purpose of the test.

10.6.9 Records and Reports

Unless otherwise prescribed by specification, the following information shall be recorded for each complete cycling series:

1. Intent of test.
2. Details of test-cycling temperature limits, soak times, and number of cycles.
3. Tray construction.
4. Extent to which chamber is loaded.
5. Extreme temperature observed at each temperature limit.
6. Last calibration date of controller and thermocouples.

7. Last date that working area of loaded chamber was profiled for thermal gradients.
8. Total and/or specification failures, type of failure, and location.

10.7 ENVIRONMENTAL THERMAL CYCLING OF PLATED PLASTICS (ASEP TP-202)

10.7.1 Scope

1. This method describes the apparatus and procedures for accurately determining the ability of a plated part to withstand specified thermal cycling routines that pertain to the environment.

2. This method simulates the temperature extremes a product could be subjected to, either in shipping or in actual use.

3. This method should serve as a guide in ranking production systems, as well as part and mold design.

4. This method yields a number, which with proper statistical techniques can be projected with defined confidence limits to predict probable product life. This projection can be valid only if product environment corresponds closely to the thermal cycling routine specified.

5. This test is a valid test for resins with a high coefficient of thermal expansion, such as ABS. It is less significant with the more stable resins that act more like metal in the environment.

10.7.2 Apparatus

The apparatus shall consist of a circulating-air heating chamber and cooling chamber sufficiently powered, insulated, and controlled so as to maintain closely the set temperature. The controller and recorder used for chamber control, calibration, and records must be accurate to 0.6°C (1°F). All points within the working area of the test chamber must remain within 2.8°C (5°F) of the set temperature when tested by the following procedures.

1. Install eight 24-gage bare thermocouples 2.5 mm (1 in) from each corner of the working area of the test chamber, plus one in the center and one at the sensing point of the chamber's controller, for a total of ten thermocouples.

2. Calibrate each thermocouple in boiling water and ice water after test setup is complete. (Also, note ASTM E-220, "Comparison Calibration of Thermocouples").

3. Load chamber with maximum quantity and type of parts to be tested.

4. Determine boundaries of control: location within chamber, maximum loading, size of sample.

10.7.3 Test Specimen

The samples (test specimens) shall be drawn at random from the quantity under evaluation. The samples may be introduced into the chamber either unmounted or mounted in a manner simulating assembly, if specified.

10.7.4 Conditioning

Since the time between plating and thermal cycling may have an effect on the thermal cycling results, tests should be run in order to determine the proper conditioning time.

10.7.5 Procedure

1. The thermal cycling routine that best stimulates the actual part environment, if not previously specified, will be selected from Table 10.4.

2. Each thermal cycle will consist of either placing the samples in a room-temperature chamber and heating the chamber to the high limit or placing the samples directly into a chamber at the high limit and performing the following:

1. One hour at the high limit.
2. One-half hour at room temperature or gradual cooling to 25 ± 5°C.
3. One hour at the low limit.
4. One-half hour at room temperature or gradual cooling to 25 ± 5°C.
5. Part examination and recording of defect type and location.

NOTE: Testing in a mounted or restrained position may affect test results.

TABLE 10.4 Recommended Cycling Routines

Cycling Routine	High Limit	Low Limit
A	85°C (185°F)	−40°C (−40°F)
B	85°C (185°F)	−29°C (−20°F)
C	74°C (165°F)	−29°C (−20°F)
D	63°C (145°F)	−29°C (−20°F)

10.7.6 Qualification Testing

Qualification testing may be initially performed to define the failure modes and percentage of failures as a function of cycle number. Ten thermal cycles should be performed (or 50% cumulative failures) so as properly to characterize the results of a system consisting of substrate, part design, mold design, molding, and plating. Sample size must be sufficient to insure adequate confidence in the estimates obtained.

10.7.7 Acceptance Testing

Acceptance testing consists of drawing a sample from each production lot or shipment and exposing this sample to one thermal cycle. The sample size and acceptance number shall be determined from the specified AQL and inspection level or from agreed-upon, defined sampling risks. A prerequisite to acceptance testing is a classification of defects, defining quality levels for defect types and location.

10.7.8 Compilation and Evaluation of Results

Parts must be carefully inspected after each complete cycle for blisters, cracks, peeling, and other defects that were not in evidence prior to thermal cycling. The severity and location of each defect should be noted, as well as the number of thermal cycles completed at the time of the inspection. Data (failures) will then be categorized in such a manner that significant characteristics can be independently evaluated. These categories may be composed of defect types, sizes, or locations, depending on the purpose of the test.

10.7.9 Records and Reports

Unless otherwise prescribed by specification, the following information shall be recorded for each complete cycling series:

1. Intent of test.
2. Details of test-cycling temperature limits, soak times, and number of cycles.
3. Tray construction.
4. Extent to which chamber is loaded.
5. Extreme temperature observed at each temperature limit.
6. Last calibration date of controller and thermocouples.
7. Last date that working area of loaded chamber was profiled for thermal gradients.
8. Total and/or specification failures, type of failure, and location.

10.8 GLACIAL ACETIC ACID TEST FOR ABS (ASEP TP-203)

10.8.1 Scope

1. This test method is for evaluating raw ABS moldings for surface stress, rubber dispersion in the surface, and internal stress.

2. This is not a quantitative test and should be used only for identifying and correcting serious stress or rubber distribution problems in raw ABS moldings.

10.8.2 Reagents

1. Anhydrous, reagent-grade glacial acetic acid is required for this test.

2. Acid should be stored in a sealer container to prevent moisture absorption from the air.

3. Do not contaminate the acid with water. Do not dip wet parts in the acid.

10.8.3 Apparatus

1. A glass or plastic container large enough to immerse the part to be tested completely is required.

2. A thermometer for checking temperature is required.

10.8.4 Procedure

1. Totally immerse the part to be tested in concentrated glacial acetic acid at 24°C ± 3°C (75°F ± 5°F) for 30 sec.

2. Immediately rinse the part in running water to remove the acetic acid, and allow it to dry at room temperature.

3. Evaluate the part for surface stress and rubber distribution.

4. After making sure the part is dry, totally immerse the part again in the concentrated glacial acetic acid at 24°C ± 3°C (75°F ± 5°F) for 2 min.

5. Immediately rinse the part in running water to remove the acetic acid, and allow it to dry at room temperature.

6. Evaluate the part for internal stress.

10.8.5 Interpretation

1. Surface stress is indicated in step 10.8.4.3 by small dense patterns appearing on the part surface. The more prevalent and dense the pattern, the higher the stress.

2. Rubber distribution in the surface is indicated by the color in step 10.8.4.3. Poor moldings will have heavy, white areas and dark, shiny areas on the same part, indicating nonuniform dispersion of the rubber phase. On an ideal molding, some surface whitening will occur uniformly over the part.

3. Cracking entirely through the thickness of the plastic in step 10.8.4.6 is an indication of very high internal stress in the part. Again, the severity of the cracking indicates the severity of the stress.

10.8.6 Reporting

1. The report should include the manufacturer of the ABS, the grade, the lot number, and the molding conditions.

2. The report should identify the degree of color variation and cracking and the locations on the part.

10.9 SOLVENT TEST FOR ABS OR MODIFIED POLYPHENYLENE OXIDE (ASEP TP-204)

10.9.1 Scope

1. This test method is intended for use by the molder to aid in determining molding cycles and conditions that minimize stress in molded parts that are to be plated.

2. This is not a quantitative test and should be used only to obtain an indication of relative stress levels in a given part.

10.9.2 Reagents

1. Anhydrous methyl ethyl ketone and acetone are required for this test.

2. Do not contaminate the solvents with water. Do not dip wet parts in the solvent.

10.9.3 Apparatus

1. A glass or plastic container large enough to immerse the part to be tested completely is required.

2. A thermometer for checking temperature is required.

10.9.4 Procedure

1. Prepare a 50-50 mixture of methyl ethyl ketone and acetone immediately prior to performing the test.

2. Immerse the parts in the solvent mixture for 15 sec at a temperature of 21°C ± 1°C (70°F ± 2°F).

3. Remove and spin dry.

4. Evaluate the part.

10.9.5 Interpretation

1. This test can show up stressed areas which may not have shown in the acetic acid test.

2. In evaluating the test results, care should be taken to ignore flaws that could have been produced by the device used to hold the part.

3. Results may vary with this test, depending on individual technique.

4. Uncontrolled variables such as actual room temperature and the rate and duration of spin drying may have an effect on the results.

10.9.6 Reporting

1. The report should include the manufacturer of the plastic, the grade, the lot number, and the molding conditions.

2. The report should identify the degree of cracking and the locations on the part.

10.10 ETHYL ACETATE TEST FOR POLYSULFONE (ASEP TP-205)

1. This test method is intended for use by the molder to aid in determining molding cycles and conditions that minimize stress in polysulfone moldings that are to be plated.

2. This is not a quantitative test and should be used only to obtain an indication of relative stress levels in a given part.

10.10.2 Reagents

1. Ethyl acetate that is at least 99% pure is required for this test.

2. The ethyl acetate should be stored in a sealed container to protect it from moisture absorption.

10.10.3 Apparatus

1. A glass or plastic container large enough to immerse the part to be tested completely is required.

2. A thermometer for checking temperature is required.

10.10.4 Procedure

1. Immerse the part in ethyl acetate for 5 to 60 sec at a temperature of 21°C ± 1°C (70°F ± 2°F).

2. Remove and let dry by evaporation.

3. A dye penetrant may be used as an aid in showing the stress cracks.

10.10.5 Interpretation

1. It is very difficult to mold stress-free polysulfone parts. Stresses brought out by this test would also be brought out by the chemicals used to condition polysulfone moldings for plating.

2. The appearance of the part after the ethyl acetate test is a very good indication of how the part will look after plating.

3. A severely stressed part will crack or break.

10.10.6 Reporting

1. The report should include the manufacturer of the plastic, the grade, the lot number, and the molding conditions.

2. The report should identify the degree of cracking and the locations on the part.

Glossary of Terms Relating to Electroplated Plastics[1]

ABS Plastic: A thermoplastic terpolymer based on acrylonitrile, butadiene, and styrene.

Acrylic Plastics: Plastics based on resins made by the polymerization of acrylic monomers, such as ethyl acrylate and methyl methacrylate.

Activation: Elimination of a passive condition on a metallic surface. Conditioning of plastic surface to initiate uniform deposition of electroless deposits.

Adhesion: The attractive force that exists between an electrodeposit and its substrate that can be measured as the force required to separate an electrodeposit and its substrate.

Anode: The electrode in electrolysis, at which negative ions are discharged, positive ions are formed, or other oxidizing reactions occur.

Antipitting Agent: An addition agent for the specific purpose of preventing gas pits in an electrodeposit.

Automatic Plating: (1) Plating in which the work being processed is automatically conveyed through successive preplate and plating baths at predetermined cycles. Called *full automatic plating*. (2) Plating in which the work being processed is conveyed automatically through only a portion of the total cycle. Called *semiautomatic plating*.

Barrel Plating or **Barrel Cleaning:** Plating or cleaning in which the work is processed in bulk in a rotating container.

Base Metal: A metal that readily oxidizes or dissolves to form ions. The opposite of *noble metal*.

[1]The majority of definitions contained herein were taken from ASTM B375-62T and ASTM D883-66T. For additional definitions relating to electroplating and plastics, see these ASTM publications.

Base Metal or **Basis Material:** Material upon which coatings are deposited. The *substrate*.

Bipolar Electrode: An electrode that is not directly connected to the power supply but is so placed in the solution between the anode and the cathode that the part nearest the anode becomes cathodic and the part nearest the cathode becomes anodic.

Blister (Plastics): A swelling of plastic, ranging from relatively small to large areas. Caused by occlusions of gas, volatile liquids, or incompatible plastic contaminant. See also *splay marks*.

Blister (Plating): A swelling of plated metal, ranging from relatively small to large areas. A bubble created between the electroplate metal and substrate or between electroplates. Usually caused by poor adhesion.

Bloom: A visible exudation or efflorescence on the surface of a plastic.

Brightener: An addition agent that leads to the formation of a bright plate or that improves the brightness of the deposit.

Bright Plating: A process that produces an electrodeposit having a high degree of specular reflectance in the as-plated condition.

Bright Plating Range: The range of current densities within which a given plating solution produces a bright plate.

Burn (Plastics): Dissolved area caused by excessive temperatures or trapped gas.

Burn (Plating): A buildup of excess plate on areas of high current density during electrolytic plating. A rough or otherwise unsatisfactory deposit produced by excessive current density.

Burn-off (of electroless plate): Removal of electroless film due to chemical attack or improper current distribution which will expose substrate.

Bus or **Bus Bar:** A rigid conducting member that carries current to the anode and cathode bars.

CASS (Copper Accelerated Acetic Acid Salt Spray Test): An accelerated test for determining the corrosion resistance of electroplated substrates, primarily applicable to copper-nickel-chromium electroplated parts. (ASTM B 368-68)

Cathode: The electrode in electrolysis at which positive ions are discharged, negative ions are formed, or other reduction actions occur.

Cleaning: The removal of contaminants, such as grease, oxides, or other foreign material from a surface.

Cold Shot or **Cold Slug:** The first material to enter an injection mold; so called because prior to entering the cavity, it is cooled below the effective molding temperature.

Composite Plate: An electrodeposit consisting of two or more layers of metal deposited successively.

Compression Molding: A method of forming objects from plastic by placing the material in a confining mold cavity and apply pressure and, usually, heat.

Conversion Coating: A coating produced by chemical or electrochemical treatment of a metallic surface that gives a superficial layer containing a compound of the metal; for example, chromate coatings on zinc and cadmium, oxide coating on steel.

Corrodkote: An accelerated test to determine the corrosion resistance of decorative chromium electroplating, primarily applicable to copper-nickel-chromium electroplated parts. (ASTM B 380-65)

Corrosion: (1) Gradual dissolution or oxidation of a metal. (2) Dissolution of anode metal by the electrochemical action in the plating cell.

Covering Power: The ability of a plating solution under a specified set of plating conditions to deposit metal on the surfaces of recesses or deep holes or other low-current-density areas. (To be distinguished from *throwing power*.) A measure of how far into a recessed area a metal will deposit.

Crazing: Fine cracks at or under the surface of a plastic or metal.

Current Density: Current (amperes) per unit area.

Current Efficiency: The proportion, usually expressed as a percentage, of the current that is effective in carrying out a specified process in accordance with Faraday's Law.

Decibel (dB): The decibel unit is a logarithmic measurement relating a variable X to a specified reference level X_0. In power, db = 10 log $P2/P1$. In voltage, db = 20 log $V2/V1$. Therefore, in power, 10 db equals 1/10 the level or 10%.

Degradation: A deleterious change in the chemical structure of a plastic.

Deionization: The removal of ions from a solution by ion exchange.

Delamination: The separation of material into layers.

Drag-in: The water or solution that adheres to or is carried by objects introduced into a bath.

Drag-out: The solution that adheres to or is carried by objects removed from a bath.

Dullness: Low reflectivity and grayer color, generally occurring in low-current-density areas and around rack clips.

Dummy or **Dummy Cathode:** A cathode in a plating solution that is not to be made use of after plating. Often used for electrolytic removal or decomposition of impurities.

Ejection Marks: Mark left by an ejector pin, blade, or sleeve that enters the cavity from the rear as the mold opens, forcing out the finished piece.

Electrode: A conductor through which current enters or leaves an electrolytic cell, at which there is a change from conduction by electrons to conduction by charged particles of matter, or vice versa.

Electrodeposition: The process of depositing a substrate upon an electrode by electrolysis.

Electroless Plating: The catalytic deposition of a metallic coating by a controlled chemical reaction that is further catalyzed by the metal or alloy being deposited.

Electrolysis: Production of chemical changes by the passage of current through an electrolyte.

Electroplating: The electrodeposition of an adherent metallic coating upon an electrode for the purpose of obtaining a surface with properties or dimensions different from those of the basis metal.

EMF (Electromotive Force): An electrical potential.

EMI: Electromagnetic interference.

Etch: A roughened surface change produced by chemical or electrochemical means.

Etchant: A chemical agent that produces a roughened surface.

Extrusion: A method whereby heated or unheated plastic forced through a shaping orifice becomes one continuously formed piece.

Fabricating: The manufacture of products from molded parts, rods, tubes, sheeting, extrusions, or other forms by appropriate operations such as punching, cutting, drilling, and tapping. Fabrication includes fastening parts together or to other parts by mechanical devices, adhesives, heat sealing, or other means.

FCC (Federal Communications Commission): This is the agency of the federal government empowered by Congress to control the use of the air waves for electromagnetic energy transmission.

Filler: A relatively inert material added to a plastic to modify its strength, permanence, working properties, or other qualities, or to lower costs.

Filter Aid: An inert, insoluble material, more or less finely divided, used as a filter medium or to assist in filtration by preventing excessive packing of the filter cake.

Flow Line: A mark or discoloration that follows the direction of the plastic flow through the molding cavity. It is usually caused by some inhomogeneity in the plastic.

Gate: The orifice through which the plastic enters the cavity in injection and transfer molding. Sometimes the gate has the same cross section as the runner leading to it; often, it is severely restricted.

Gate Blushing: A surface defect, usually a dulling of the surface or splay marks around the area of the gate.

Gate Mark: Defect on plastic where gate has been removed.

Gigahertz (GHz): Equals 1 million Hertz.

Hertz (Hz): Hertz is the unit used to define frequency. One Hertz equals one cycle per second.

Hydrophilic: (1) Tending to attract water. (2) Having an affinity for water.

Hydrophobic: (1) Tending to repel water. (2) Lacking affinity for water.

Immersion Plate: A metallic deposit produced by a displacement reaction in which one metal displaces another from solution; for example, $Fe + Cu^{++} \rightarrow Cu + Fe^{++}$.

Injection Molding: A method of forming plastic objects from granular or powdered plastics by fusing the plastic in a chamber with heat and pressure and then forcing part of the mass into a cooler chamber, where it solidifies.

Kilohertz (kHz): Equals 1000 Hertz.

Leveling Action: The ability of a plating solution to produce a plated surface smoother than that of the substrate.

Megahertz (MHz): Equals 1 million Hertz.

Metalizing: (1) The application of an electrically conductive metallic layer to the surface of nonconductors. (2) The application of metallic coatings by nonelectrolytic procedures such as spraying of molten metal, deposition from the vapor phase, or catalytic electroless deposition from a solution of metallic salts.

Mil: One-thousandth of an inch; .001 in. = 25.4 μm.

Molding: See *compression molding, injection molding, transfer molding*.

Mold Release Agent: A material applied to the surface of the mold which makes the removal of the molded part easier.

Monomer: A relatively simple compound that can react to form a polymer.

Necking: The localized reduction in cross section that may occur in a material under tensile stress.

Nylon Plastic: A plastic based on resins composed principally of a long-chain synthetic polymeric amide that has recurring amide groups as an integral part of the main polymer chain.

Off-Color: Mismatch to a standard color.

Parting Line: Mark on a molding or casting where halves of a mold met in closing.

Passivity: The condition of a metal that retards its normal reaction in a specified environment and associated with the assumption of a potential more noble than its normal potential.

Peeling (Plastics): Delamination; the separation of the layers in a laminate caused by adhesion failure.

Peeling (Plating): Detachment or partial detachment of an electrodeposited coating from the plastic substrate or between layers of electrodeposited metal.

Peel Test: A test for plated plastic in which the force required to separate the metal from the plastic substrate is measured.

Phenolic Plastics: Plastics based on resins made by the condensation of phenols, such as phenol and cresol, with aldehydes, such as formaldehyde.

Pinhole: Very small hole in a plastic or in a metal deposit.

Pit (Plastics): An unwanted, small depression or cavity. (Presence of pits in the mold causes projections that result in roughness of the molded part. Gas or water in the plastic melt may cause pits in the plastic part.)

Plastic: A material that contains as an essential ingredient an organic substance of large molecular weight, is solid in its finished state, and, at some stage in its manufacture or in its processing into finished articles, can be shaped by flow.

Plasticizer: A material incorporated in a plastic to increase its workability and its flexibility or distensibility.

Plastisol: A mixture of resins, plasticizers, and other minor additives, such as pigments, that can be converted to a continuous film by the application of heat. Distinct from baking enamels and other substances in that substantially all the original mixture becomes a part of the film; there is no significant evaporation of solvent. The films are usually much thicker than obtainable from coatings that depend on the evaporation of a volatile solvent.

Plating Range: The current-density range over which a satisfactory electroplate can be deposited.

Polyethylene: A plastic or resin prepared by the polymerization of ethylene as essentially the sole monomer.

Polymer: A compound formed by the reaction of simple molecules having functional groups that permit their combination to proceed to high molecular weights under suitable conditions. Polymers may be formed by addition polymerization (through unsaturated bonds) or polycondensation (splitting out water, etc.). When two or more monomers are involved, the product is called a *copolymer*.

Polymerization: A chemical reaction in which the molecules of a monomer are linked together to form large molecules whose molecular weight is a multiple of that of the original substrate. When two or more monomers are involved, the process is called *copolymerization* or *heteropolymerization*.

Polypropylene: A plastic or resin prepared by the polymerization of propylene as essentially the sole monomer.

Polystyrene: A plastic based on a resin made by polymerization of styrene as the sole monomer.

Porosity: Presence of numerous small voids.

Rack (Plating): A frame for suspending and carrying current to articles during plating and related operations.

Rectifier: A device that converts alternating into direct current by virtue of a characteristic that permits appreciable flow of current in only one direction.

Regrind: See *reworked material*.

Reinforced Plastic: A plastic with some strength properties greatly superior to those of the base resin, resulting from the presence of high-strength fillers compounded in the composition.

Resin: A solid, semisolid, or pseudosolid organic material that has an indefinite and often high molecular weight, exhibits a tendency to flow when subjected to stress, usually has a softening or melting range, and usually fractures conchoidally.

Resist Paint: A paint applied to plastic prior to plating that prevents plate deposition in those areas painted.

Reworked Material: A plastic material that has been reprocessed after having been previously processed by molding extrusion, etc., in a fabricator's plant. Also known as *regrind*.

Robber: See *thief*.

Runner: (1) The secondary feed channel that runs from the inner end of the sprue in an injection mold or transfer mold to the cavity gate. (2) The piece formed in a secondary feed channel or runner.

Scratch or **Nick:** A break or mar on the surface of the electroplated finish or in the plastic substrate.

Sheet Resistance: The electrical resistance between two electrodes as measured by ASTM method F-390. This method is easy to perform and gives some preliminary indication of shielding effectiveness but is not conclusive except in very gross terms. Units are ohms per square and a value of 1 ohm per square or less is generally considered necessary for shielding applications.

Shelf Roughness: Roughness on a horizontal surface where insoluble contaminants have settled on parts during the plating operation.

Shield: A nonconducting medium for altering the current distribution on an anode or a cathode.

Short Shot: A piece that results from an incompletely filled mold.

Shrink Mark: A depression in the surface of a molded material where it has retracted from the mold.

Sink Mark: A shallow depression or dimple on the surface of an injection-molded part due to retracting of the surface following local internal shrinkage after the gate seals. May indicate need for design or processing change.

Splay Marks: A fan-shaped array of bubbles or streaks, usually near the gate area but also occurring elsewhere. Caused by occlusions of gas or volatile liquids. See also *blister (plastics)*.

Sprue: (1) The primary feed channel that runs from the outer face of an injection or transfer mold to the mold gate in a single-cavity mold or to the runners in a multiple-cavity mold. (2) The piece formed in a primary feed channel or sprue.

Stain: A discoloration of a surface.

Stardusting: An extremely fine form of roughness on the surface of a metallized part.

Stray Current: Current through paths other than the intended circuit, such as through heating coils or the tank.

Stress Crack: External or internal cracks in a plastic or electrodeposit caused by tensile stresses less than that of its short-time mechanical strength.

Strike: (1) A thin film of metal to be followed by other coatings. (2) A solution used to deposit a strike. (3) To plate for a short time, usually at a high initial current density. In plating on plastics, the term *strike* is often used to designate thin deposits obtained at low current densities to prevent burn-off of electroless deposit.

Styrene Plastics: Plastics based on resins made by the polymerization of styrene or copolymerization of styrene with other unsaturated compounds, the styrene being in greatest amount by weight.

Substrate: See *basis metal*.

Thermal Cycle Test: A test for plated plastics designated to determine the ability of the composite to withstand severe temperature changes. May be used as an adhesion test.

Thermal Degradation: A deleterious change in the chemical structure of the plastic due to local overheating.

Thermoforming: Forming with the aid of heat.

Thermoplastic: Capable of being repeatedly softened by increase of temperature and hardened by decrease of temperature.

Thermoset: A plastic which, when cured by application of heat or by chemical means, changes into a substantially infusible and insoluble product when cured under application of heat or by chemical means.

Thief: An auxiliary cathode so placed as to divert to itself some current from portions of the work which would otherwise receive too high a current density. Also called *robber*.

Throwing Power: The improvement of the (usually metal) coating distribution over the primary current distribution on an electrode (usually a cathode) in a given solution, under specified conditions. The term may also be used for anodic processes for which the definition is analogous.

Transfer Molding: A method of forming plastic objects from granular, powdered, or preformed plastics by fusing the plastic in a chamber with heat and then forcing essentially the whole mass into a hot chamber, where it solidifies.

Vacuum Forming: A forming process in which a heated plastic sheet is changed to a desired shape by causing it to flow by reducing the air pressure on one side of the sheet.

Vinyl Plastics: Plastics based on resins made from monomers containing the vinyl group, $CH_2 = CH -$.

Void (Plastic): An apparent gas bubble entrapped in an otherwise solid mass.

Void (Plating): An area in which a part of the substrate is visible after plating.

Warp: Dimensional distortion in a plastic object after molding or other fabrication.

Warpage: Distortion of a plastic piece due to internal stresses.

Water Break: The appearance of a discontinuous film of water on a surface, signifying nonuniform wetting and usually associated with surface contamination.

Weld Mark or **Weld Marker Line:** A mark on a molded plastic piece made by the meeting of two flow fronts during the molding operation.

Suggested Readings

ASTM designation B-368-68.

ASTM standard thermal-cycle test, ASTM designation B-553.

Abu-Isa, I.A., "Iodine Treatment of Nylon: Effect on Metal Plating of the Polymer," *J. App. Polym. Sci.*, 15, 1971, 2865.

Abu-Isa, I.A., US Patent No. 3695917.

Armond, V.J. and Atkinson, J.R., "Chromic Acid as an Etchant for Bulk Polypropylene and Its Use to Study; (1) Nitric Acid Attack on Polypropylene, (2) Cracks in Polypropylene Induced by Tensile Stress," *J. Mater. Sci.*, 4, 1969, 509.

Arrowsmith, D.J., "Aspects of Adhesion," *Prod. Finish.*, 24, 1971, (1), 40.

Atkinson, E.B. et al., "Metal Plating of ABS and Other Polymers," *Trans. J. Plast. Inst.*, 35, 1967, 549.

Beacom, S.E. and Wedel, R.G., "SEM Microscopy of Plastic Surfaces as Related to Metal Deposition," *J. Colloid Interface Sci.*, 34, 1970, 375.

Bruce, J.G. and Roobal, N.R., *Plating*, 60, 1973, 491.

Carter, V.E., "The Assessment of Thermal Stability and Corrosion Resistance of Electroplated Coatings on ABS Plastics," *Trans. Inst. Met. Finish.*, 46, 1968, 49.

Carter, V.E., "The Outdoor Corrosion Performance of Plated ABS Plastics," *Trans. Inst. Met. Finish.*, 53, 1975, 61.

Carter, V.E., "The Validity of Laboratory Tests on Plated ABS for Assessing Service Behavior," *Trans. Inst. Met. Finish.*, 49, 1971, 29.

Chow, S.L., "The Roles of UV Light in the Inhibition of Electroless Deposition," *et al., J. Electrochem. Soc.*, 119, 1972, 1013.

Cohen, R.L., "Mossbauer Study of Tin (II) Sensitizer Deposit on Kaptan," *et al., J. Electrochem. Soc.*, 118, 1971, 2042.

Cohen, R.L. and West, K.W., "Characterization of Tin Palladium Sols.," *Chem. Phys. Lett.*, 16, 1972, 128.

Cohen, R.L. and West, K.W., "Generation and Stabilizing Processes in Tin Palladium Sols and Palladium Sol Sensitizers," *J. Electrochem. Soc.*, 120, 1973, 502.

Cohen, R.L. and West, K.W., "Solution Chemistry and Colloid Formation in the Tin Chloride Sensitizing Process," *J. Electrochem. Soc.*, 119, 1972, 433.

Coombes, R.L., "Electroless Copper Preplating on ABS Plastics," *Plating*, 57, 1970, 675.

Crouch, P.C., "Effect of Nickel Thickness and Copper Undercoats on Performance of Plated ABS Plastics," *Trans. Inst. Met. Finish.*, 49, 1971, 141.

Crouch, P.C., "Performance of Plated ABS Plastic in Thermal Cycling and Corrosion Tests," *Trans. Inst. Finish.*, 49, 1971, 141.

Crouch, P.C., "The Performance of Metal ABS Plastics During Atmospheric Exposure," *Trans. Inst. Met. Finish.*, 51, 1973, 117; also 52, 1973, 149.

D'Amico, J.F., "Selective Electroless Metal Deposition Using Patterned Photo-Oxidation of Su II Sensitized Substrates," *et al., J. Electrochem. Soc.*, 118, 1971, 1695.

Dennis, J.K., *Int. Metall. Rev.*, 17, 1972, 43.

Dennis, J.K. and Tipping, P., "A Study of Factors Affecting Peel Adhesion of Electroplated Coatings on ABS and Polypropylene," *Electroplating*, 27, 1974, (11), 9.

Dennis, J.K. and Tipping, P., "Corrosion Resistance of Nickel and Chromium Coating," *Trans. Inst. Met. Finish.*, 52, 1974, 5 and 133.

DiBari, G.A. Paper presented at the 8th Annual Meeting of the American Society of Electroplated Plastics, Palm Springs, CA, 1975.

DiBari, G.A., "Plated Plastics Corrosion Performance and Speculations," *Plating*, 60, 1973, 1252.

Ebneth, H. and Klimaschewski, W., *Metal Finish. J.*, 1969, 317.

Ebneth, H., et al., *Trans. J. Plastic. Inst.*, 35, 1967, 543.

Ellis, P.A.M., "The Importance of Molding Conditions to the Quality of Electroplated ABS Plastics," *Trans. J. Plast. Inst.*, 35, 1967, 537.

Feldstein, N., "Electron Microscope Studies of an Improved Sensitizer Solution," *et al., J. Electrochem. Soc.*, 120, 1973, 875.

Feldstein, N., "Reliability in Printed Circuitry Metalization: A Case for Improved Catalyzing Systems," *Plating*, 60, 1973, 611.

Feldstein, N. and Weiner, J.A., "Some Aspects of the Chemistry of Tin Sensitizing Solutions," *et al., J. Electrochem. Soc.*, 119, 1972, 669.

Feldstein, N. and Weiner, J.A., "Some Aspects of the Chemistry of Tin Sensitizing Solutions," *et al., Plating*, 119, 1972, 1486.

Feldstein, N. and Weiner, J.A., "Surface Characterization of Sensitized Activated Teflon," *J. Electrochem. Soc.*, 120, 1973, 475.

Fitchmum, D.R., "Electroplating on Crystalline Polypropylene II Injection Molding and Adhesion," *et al., J. Appl. Polym. Sci.*, 14, 1970, 2441.

Flowers, E.B., "Plating Adhesion Problems in the Manufacture of Printed Wiring Boards—Types, Causes, and Corrective Action—A Review," *Plating*, 59, 1972, 953.

Frieze, A.S. and Weil, R., *Electron Microscopy*, ed. R. Uzeda, Vol. 1, Tokyo: Maruzen Company, Ltd., 1966.

Gentle, D.F., "Metalization of Plastics—Process Selection for Automotive Use," *Electroplating*, 24, 1971, (2), 12.

Ghorashi, H.M., "Effect of Mold Temperatures on Etching of Polypropylene," *Plating*, 63, 1976, 52.

Goldie, W., *Met. Finish.*, 62, 1964, (12), 50.

Goldie, W., *Metallic Coating of Plastics*, Vols. 1 and 2, Middlesex, England: Electrochemical Publications, Ltd, 1968.

Grunwald, J.J. and D'Ottavio, E.D., US Patent No. 3716392.

Hamilton, J.F. and Logel, P.C., "Catalysis of Electroless Nickel Deposition by Small Palladium Nickels," *J. Catal.*, 29, 1973, 253.

Hepfer, I.C., "Electroplated Plastics: Environmental and Accelerated Corrosion Studies," *et al., Plating*, 58, 1971, 333.

Hepfer, I.C., *Plating*, 55, 1968, 584.

Heyman, K., "Electroplating of Plastics in Theory and Practice," *et al., Angew Chem.*, 9, 1970, 425.

Heyman, K., "Electroplating of Plastics in Theory and Practice," *et al., Prod. Finish.*, 24, 1971, (10), 20.

Heyman, K., Proc., Int. Conf. on Protection Against Corrosion by Metal Finishing, Basel, 1966, ed. N. Ibl. et al., Zurich, 1967, Forster Verlag AG.

Heyman, K. and Woldt, G., *Galvanotechnik*, 61, 1970, 221; and with W. Riedel, *Angew Chem (Eng.)*, 9, 1970, 425.

Hurley, J.L. and Chart, J.E., *Modern Electroplating*, ed. F.A. Lowenheim, 3rd ed., New York: John Wiley, 1974.

Hurley, J.L. and Chart, J.E., *Plating*, 62, 1975, 127.

Innes, W.P., "Chromium Plated ABS and Polypropylene Plastics: Performance in Typical Tests," *et al., Plating*, 56, 1969, 51.

Innes, W.P., "Thermal Cycle Performance of Chromium Plated ABS Headlamp Bezel," *et al., Plating*, 58, 1971, 1002.

Innes, W.P., "The Use of Plated Polypropylene for Exterior Automotive Decorative Trim," *et al., Plating*, 57, 1970, 1006.

Janssen, C.J.F.G., *et al., Plating*, 58, 1971, 42.

Kadison, L.A. and Maguire, E., US Patent No. 3668130.

Kato, K., "ABS Mouldings for Electroplating an Electron Microscope Study," *Polymer*, 8, 1967, 33.

Kato, K., "Electron Microscopic Studies on the Etching of ABS Mouldings for Electroplating," *Polymer*, 9, 1968, 419.

Kato, K., "Molding Anisotrophy in ABS Polymers as Revealed by Electronmicroscopy," *Polymer*, 9, 1968, 225; also, *J. Electron Microsc.*, 17, 1968, 23.

Kern, O.A. and Porter, C.L., SAE preprint 1970, No. 700017.

Ladney, Jr., M., US Patent No. 3725216.

Levy, J., *Tech. Proc. Am. Electroplat. Soc.*, 51, 1964, 139.

Logie, G.R. and Rantell, A., "The Adhesion of Electroless Metal Deposits to ABS and Other Polymers," *Trans. Inst. Met. Finish.*, 46, 1968, 91.

Logie, G.R. and Rantell, A., *Trans. Inst. Met. Finish.*, 46, 1968, 91.

Maguire, E. and Kadison, L.A., US Patent No. 3671289.

Mason, B., "Chemicals for Plating Through Holes by Additive Techniques," *Electroplating*, 25, 1972, (6), 14.

Matsunaga, M., "Adhesion of Electrodeposits to Plastics on Electron Microscopic Investigation," *et al., Met. Fin.*, 66, 1968, (11), 80.

Matsunaga, M. and Hagiuda, Y., "Mechanical Properties of Electroplated ABS Plastics," *Met. Finish.*, 69, 1971, (4), 36.

Matsunaga, M., *et al., Met. Finish.*, 66, 1968, (11), 80.

McBride, D.G. and Vlasak, G.P., *J. Electrochem. Soc.*, 118, 1971, 2055.

McDermott, J., *Plating of Plastics with Metals*, Park Ridge, New Jersey: Noyes Data Corporation, 1974.

McGregor, A. and Perrins, L.E., "Some Factors Influencing the Adhesions of Electroplated Copper to Polypropylene," *Plast. Polym.*, 38, 1970, 193.

Melka, J.P., "Effects of Silver Nitrate Activation on Electroless Copper Deposit," *Plating*, 58, 1971, 1085.

Miller, G.T., US Patent No. 3556956, 1971.

Miller, G.T., US Patent No. 3510327, US Patent No. 3556956.

Minjer, C.H. de and Bloom, P.F.J.v.d., "The Nicklation with S_nU_2–$PdCl_2$ Solutions of Glass Before Electroless Plating," *J. Electrochem. Soc.*, 120, 1973, 1644.

Morton, G.H. and Baier, F.L., *Soc. Plast. Eng. Tech. Pap.*, 13, 1967, 888.

Morton, J.P. and Schlesinger, M., *J. Electrochem. Soc.*, 115, 1968, 15.

Munson, R.H. and Melford, A.M., "Stallographic Examination Forces in Electro-Plated Polypropylene," SAE preprint, No. 690095, 1969.

Nakamura, Y., *et al., J. App. Polym. Sci.*, 16, 1972, 2727.

Narcus, H., *Metalizing of Plastics*, New York: Reinhold, 1960.

Narcus, H., "Plating on Plastics Practical Plant Operation and Trouble Shooting," *Plating*, 55, 1968, 816.

Pearlstein, F., *Modern Electroplating*, ed. F.A. Lowenheim, 3rd ed., New York: John Wiley, 1974, p. 710.

Perrins, L.E. and Pettett, K., "Mechanism for the Adhesion of Electroplate to Propylene Copolymer," *Trans. Inst. Met. Finish.*, 51, 1973, 145; 50, 1972, 30; also, *Plast. Polym.*, 39, 1971, 391.

Petrow, C., *et al., Kunststoffe*, 59, 1969, 229.

Poppe, W., "Interpretations of Adhesive Forces in Electroplated Polypropylene," *J. Adh.*, 2, 1970, 144.

Rantell, A., "The Effect of Surface Roughness and Substrate Toughness on the Peel Value Obtained with Electroplated Plastics," *Trans. Inst. Met. Finish.*, 50, 1972, 146.

Rantell, A., "The Influence to Surface Chemistry on the Adhesion of Copper Deposits on Plastic Substrates," *Trans. Inst. Met. Finish.*, 47, 1969, 197.

Rantell, A., *Trans. Inst. Met. Finish.*, 48, 1970, 191.

Rantell, A. and Holtzman, A., "Activations of ABS Using Mixed Catalysts," *Met. Finish.*, 71, 1973, (9), 47.

Rantell, A. and Holtzman, A., "Mechanism of Activator of Polymer Surfaces by Mixed Stannous Chloride/Palladium Chloride Catalysts," *Trans. Inst. Met. Finish.*, 51, 1973, (2), 62.

Rantell, A. and Holtzman, A., "Mixed Stannous Chloride Palladium Chloride Activator—A Study of their Formation and Nature," *Plating*, 61, 1974, 326.

Rantell, A. and Holtzman, A., "The Role of Accelerators Prior to Electroless Plating of ABS Plastic," *Trans. Inst. Met. Finish.*, 52, 1974, (2), 31.

Rantell, A. and Holtzman, A., "The Role of Rinsing During the Activator of ABS Plastics Using Mixed $S_nCl_2/PdCl_2$ Catalysts," *Electroplating*, 27, 1974, (2), 15.

Rempel, D., "Experiences with Electroplating Plastic Foils," *Electroplating*, 23, 1970, (2), 27.

Rempel, D., et al., "Thermoplastic of ABS Graft Polymers Capable of Being Electroplated," *Prod. Finish.*, 24, 1971, (2), 22.

Roobal, N.R. and Delano, T., "Designing for Plated Polypropylene Parts, *SPE J.*, 27, 1971, (8), 70.

Sard, R., *J. Electrochem. Soc.*, 117, 1970, 864.

Saubestre, E.B., *Met Finish.*, 60, 1962, (6), 67; also 60, 1962, (7), 49.

Saubestre, E.B., *Modern Electroplating*, ed. F.A. Lowenheim, 3rd ed., New York: John Wiley, 1974, p. 636.

Saubestre, E.B., *Plating*, 52, 1965, 982.

Saubestre, E.B., *Proc. Am. Electroplat. Soc.*, 46, 1959, 264.

Saubestre, E.B., "The Nature of the Corrosion of Electroplated Deposits on Plastics," *et al., Electrochem. Techn.*, 6, 1968, 341.

Saubestre, E.B. and Hajdu, J.H., US Patent No. 3471320.

Schwartz, E. and Hartley, J.L., "The Acetic Acid Test–Useful Tool for Stress Control and Plating Quality," *SPE J.*, 23, 1967, (3), 110.

Sharpe, D.J., "Photoselective Metal Deposition," *Plating*, 58, 1971, 786.
Shipley, Jr., C.R., US Patent No. 3011920.
Smith, R.R. and Lewis, T.D., *Electroplating*, 21, 1968, 44.
Spencer, L.F., "Electroless Nickel Plating–A Review," *Met. Finish.*, 73, 1975, (1), 38; also 72, 1974, (10), 35; and 72, 1974, (12), 58.
Steen, S.S., "Molding ABS for Electroplating," *SPE J.*, 24, 1968, (10), 77.
Such, T.E. and Baldwin, C., "The Adhesion Testing of Electroplated Coatings on ABS Moldings," *Trans. J. Plast. Inst.*, 35, 1967, 553.
Tabei, H. et al., *J. Electrochem. Soc.*, 121, 1974, 67.
Tipping, P. and Dennis, J.K., "Tensile Strength, Ductility, and Impact Strength of Plated ABS Polypropylene," *Trans. Inst. Met. Finish.*, 53, 1975, (1), 1.
Tison, R.P., *Plating*, 60, 1973, 339.
Tison, R.P., "Stress/Strain Analysis in Plated Plastics," *Plating*, 60, 1973, 47.
Tison, R.P., unpublished calculations.
Wedel, R.G., "A Mechanism for Corrosion-Associated Blisters on Plated Plastics," *Plating*, 62, 1975, 235.
Wedel, R.G., "Corrosion-Associated Blistering of Plated Plastics," *Plating*, 62, 1975, 40.
Wedel, R.G., "The Effects of Surface Porosity on the Plateability and Adhesion of Metals to Nonconductors," *Plating*, 58, 1971, 225.
Wedel, R.G., *Plating*, 62, 1975, 40.
Wein, S., "Silver Films," PB 11236, OTS, 1953, US Dept. Commerce, Washington, D.C.
Weston, N.E., "Inorganic Fiber Reinforcement of Plated ABS," *Soc. Plastics Eng. Tech. Pap.*, 18, 1972, 171.
Weibusch, K., "Bond Strength of Metal and Plastic," *Ind. Finish*, Nov. 1969, p. 27.
Weibusch, K., "Influence of Molding Conditions and Pretreatment on the Properties," *Prod. Finish*, 19, 1966, (11), 44; also, *Kunststoffe*, 56, 1966, 773.
Wiggle, R.R., "The Mechanism of Adhesion Failure of Plated Plastics in Corrosive Environments," *et al., J. Electrochem. Soc.*, 118, 1971, 158.
Woldt, G., "Experimental Results Obtained in the Electroplating of Plastics with the Use of the Scanning Electron Microscope," *Trans. Inst. Met. Finish.*, 47, 1969, (5), 236.
Zahn, E. and Wiebusch, K., "Experiments on the Plating of ABS Substrates," *Kunststoffe*, 56, 1966, 773.
Zappala, J.F., "Molding and Quality Analysis of ABS Substrates," *Plating*, 56, 1969, 459.